Lecture Notes in Physics

T0192277

The Lecture Notes in Physics

The series Lecture Notes in Physics (LNP), founded in 1969, reports new developments in physics research and teaching – quickly and informally, but with a high quality and the explicit aim to summarize and communicate current knowledge in an accessible way. Books published in this series are conceived as bridging material between advanced graduate textbooks and the forefront of research and to serve three purposes:

- to be a compact and modern up-to-date source of reference on a well-defined topic

- to serve as an accessible introduction to the field to postgraduate students and nonspecialist researchers from related areas

- to be a source of advanced teaching material for specialized seminars, courses and schools

Both monographs and multi-author volumes will be considered for publication. Edited volumes should, however, consist of a very limited number of contributions only. Proceedings will not be considered for LNP.

Volumes published in LNP are disseminated both in print and in electronic formats, the electronic archive being available at springerlink.com. The series content is indexed, abstracted and referenced by many abstracting and information services, bibliographic networks, subscription agencies, library networks, and consortia.

Proposals should be sent to a member of the Editorial Board, or directly to the managing editor at Springer:

Christian Caron
Springer Heidelberg
Physics Editorial Department I
Tiergartenstrasse 17
69121 Heidelberg / Germany
christian.caron@springer.com

F. Strocchi

Symmetry Breaking

Second Edition

 Springer

Author

Franco Strocchi
Scuola Normale Superiore
Classe di Scienze
Piazza dei Cavalieri 7
56100 Pisa, Italy

F. Strocchi, *Symmetry Breaking*, Lect. Notes Phys. 732 (Springer, Berlin Heidelberg 2008), DOI 10.1007/978-3-540-73593-9

This first edition of this book was originally published as Vol. *643* in the series *Lecture Notes in Physics*

ISSN 0075-8450
ISBN 978-3-642-09273-2   e-ISBN 978-3-540-73593-9

Springer is a part of Springer Science+Business Media
springer.com
© Springer-Verlag Berlin Heidelberg 2008
Softcover reprint of the hardcover 2nd edition 2008

Cover design: eStudio Calamar S.L., F. Steinen-Broo, Pau/Girona, Spain

Preface

The main motivation for such lecture notes is the importance of the concept and mechanism of spontaneous symmetry breaking in modern theoretical physics and the relevance of a textbook exposition at the graduate student level beyond the oversimplified (non-rigorous) treatments, often confined to specific models. One of the main points is to emphasize that the radical loss of symmetric behaviour requires both the existence of non-symmetric ground states and the infinite extension of the system.

The first Part on SYMMETRY BREAKING IN CLASSICAL SYSTEMS is devoted to the mathematical understanding of spontaneous symmetry breaking on the basis of classical field theory. The main points, which do not seem to appear in textbooks, are the following.

i) **Existence of disjoint Hilbert space sectors**, stable under time evolution in the set of solutions of the classical (non-linear) field equations. They are the strict analogs of the different phases of statistical mechanical systems and/or of the inequivalent representations of local field algebras in quantum field theory (QFT). As in QFT, such structures rely on the concepts of locality (or localization) and stability, (see Chap. 5), with emphasis on the physical motivations of the mathematical concepts; such structures have the physical meaning of *disjoint physical worlds*, disjoint *phases* etc. which can be associated to a given non-linear field equation. The result of Theorem 5.2 may be regarded as a generalization of the criterium of stability to infinite dimensional systems and it links such stability to elliptic problems in $\mathbf{R}^n$ with non-trivial boundary conditions at infinity (Appendix E).

ii) Such structures allow to reconcile the classical **Noether theorem** with spontaneous **symmetry breaking**, through a discussion of a mechanism which accounts for (and explains) the breaking of the symmetry group (of the equations of motion), in a given Hilbert space sector $\mathcal{H}$, down to the subgroup which leaves $\mathcal{H}$ stable (Theorem 7.1).

iii) The classical counterpart of the **Goldstone theorem** is proved in Chap. 9, which improves and partly corrects the heuristic perturbative arguments of the literature.

The presentation emphasizes the general ideas (implemented in explicit examples) without indulging on the technical details, but also without derogating from the mathematical soundness of the statements.

The second Part on SYMMETRY BREAKING IN QUANTUM SYSTEMS tries to offer a presentation of the subject, which should be more mathematically sounded and convincing than the popular accounts, but not too technical. The first chapters are devoted to the general structures which arise in the quantum description of infinitely extended systems with emphasis on the physical basis of *locality, asymptotic abelianess* and *cluster property* and their mutual relations, leading to a characterization of the **pure phases**.

Criteria of spontaneous symmetry breaking are discussed in Chap. 8 along the lines of Wightman lectures at Coral Gables and their effectiveness and differences are explicitly worked out and checked in the Ising model. The Bogoliubov strategy is shown to provide a simple rigorous control of spontaneous symmetry breaking in the free Bose gas as a possible alternative to Cannon and Bratelli-Robinson treatment.

The **Goldstone theorem** is critically discussed in Chap. 15, especially for non-relativistic systems or more generally for systems with long range delocalization. Such analysis, which does not seem to appear in textbooks, provides a non-perturbative explanation of *symmetry breaking with energy gap* in **non-relativistic Coulomb systems** and in the **Higgs phenomenon** and in our opinion puts in a more convincing and rigorous perspective the analogies proposed by Anderson. The Swieca conjecture about the role of the potential fall off is checked by a perturbative expansion in time. Such an expansion also supports the condition of integrability of the charge density commutators, which seems to be overlooked in the standard treatments and plays a crucial role for the energy spectrum of the Goldstone bosons. As a result of such an explicit analysis, the *critical decay of the potential* for allowing "massive" Goldstone bosons turns out to be that of the Coulomb potential, rather than the one power faster decay predicted by Swieca condition.

The *non-zero temperature* version of the *Goldstone theorem*, discussed in Chap. 16, corrects some wrong conclusions of the literature. An extension of the Goldstone theorem to non-symmetric Hamiltonians is discussed in Chap. 18 with the derivation of non-trivial (non-perturbative) information on the energy gap of the modified Goldstone spectrum.

The symmetry breaking in gauge theories, in particular the **Higgs phenomenon** which is at the basis of the standard model of elementary particles, is analyzed in Chapter 19. The problems of the perturbative explanation of the evasion of the Goldstone theorem are pointed out and a non-perturbative account is presented. In the local renormalizable gauges, the absence of physical Goldstone bosons follows from the Gauss law constraint or subsidiary condition on the physical states. In the Coulomb gauge, the full Higgs phenomenon is explained by the failure of relative locality between the current and the Higgs field, by exactly the same mechanism discussed in Chap. 15 for the non-relativistic Coulomb systems; in particular, the Goldstone spectrum is shown to be given by the Fourier spectrum of the two point function of the vector boson field, which cannot have a $\delta(k^2)$ contribution, since otherwise the symmetry would not be broken.

The chapters marked with a * can be skipped in a first reading.

The second edition differs from the first by the correction of some misprints, by an improved discussion of some relevant points and by a significantly expanded and more detailed discussion of symmetry breaking in gauge theories.

Contents

Symmetry Breaking in Classical Systems

Introduction to Part I

These notes essentially reproduce lectures given at the International School for Advanced Studies (Trieste) and at the Scuola Normale Superiore (Pisa) on various occasions. The scope of the short series of lectures, typically a fraction of a one-semester course, was to explain on general grounds, also to mathematicians, the phenomenon of Spontaneous Symmetry Breaking (SSB), a mechanism which seems at the basis of most of the recent developments in theoretical physics (from Statistical Mechanics to Many-Body theory and to Elementary Particle theory).

Besides its extraordinary success, the idea of SSB also deserves being discussed because of its innovative philosophical content, and in our opinion it should be part of the background knowledge for mathematical and theoretical physics students, especially those who are interested in questions of principle and in general mathematical structures.

By the general wisdom of Classical Mechanics, codified in the classical Noether theorem, one learns that the symmetries of the Hamiltonian or of the Lagrangean are automatically symmetries of the physical system described by it, which does not mean that the (equilibrium) solutions are symmetric, but rather that the symmetry transformation commutes with time evolution and hence is a symmetry of the physical behaviour of the system. This belief therefore precludes the possibility of describing systems with different dynamical properties in terms of the same Hamiltonian. The realization that this obstruction does not *a priori* exist and that one may unify the description of apparently different systems in terms of a single Hamiltonian and account for the different behaviours by the mechanism of SSB, is a real revolution in the way of thinking in terms of symmetries and corresponding properties of physical systems. It is, in fact, non-trivial to understand how the conclusions of the Noether theorem can be evaded and how a symmetry of the dynamics cannot be realized as a mapping of the physical configurations of the system, which commutes with the time evolution.

The standard folklore explanations of SSB, which one often finds in the literature, is partly misleading, because it does not emphasize the crucial ingredient underlying the phenomenon, namely the need of infinite degrees of freedom. Despite the many popular accounts, the phenomenon of SSB is deep and subtle and it is not without reasons that it has been fully understood only in recent times. The standard cheap explanation identifies the phenomenon with the existence of a degenerate ground (or equilibrium) state, unstable under the symmetry operation, (*ground state asymmetry*), a feature often present even in simple mechanical models (as for example a particle on a plane, each point of which defines a ground state unstable under translations), but which is usually not accompanied by a non-symmetric behaviour.

As it will be discussed in these lectures, the phenomenon of spontaneous symmetry breaking in the radical sense of non-symmetric behaviour is rather related to the fact that, for non-linear infinitely extended systems (therefore involving infinite degrees of freedom), the solutions of the dynamical problem generically fall into classes or "islands" or "phases", stable under time evolution and characterized by the same behaviour at infinity of the corresponding solutions. Since all physically realizable operations have an inevitable localization in space they cannot change such a behaviour at infinity and therefore starting from the configurations of a given islands one cannot reach the configurations of a different island by physically realizable modifications. The different islands can then be interpreted as describing physically disjoint realizations or *different phases*, or *disjoint physical worlds* associated with the given dynamics.

The spontaneous breaking of a symmetry (of the dynamics) in a given phase or physical world can then be explained as the result of the instability of the given island under the symmetry operation. In fact, in this case one cannot realize the symmetry within the given island, namely one cannot associate with each configuration the one obtained by the symmetry operation.

The existence of such structures is not obvious and in general it involves a mathematical control of the non-linear time evolution of systems with infinite degrees of freedom and the mathematical formalization of the concept of physical disjointness of different islands. For quantum systems, where the mathematical basis of SSB has mostly been discussed, the physical disjointness has been ascribed to the existence of inequivalent representations of the algebra of *local* observables.

The scope of Part I of these lectures is to discuss the general mechanism of SSB within the framework of classical dynamical systems, so that no specific knowledge of quantum mechanics of infinite systems is needed and the message may also be suitable for mathematical students. More specifically, the discussion will be based on the mathematical control of the non-linear evolution of classical fields, with *locally* square integrable initial data which may possibly have non-vanishing limits at infinity.

The mathematical formalization of physical disjointness relies on the constraint of essential localization in space of any physically realizable operation. One can in fact show that an island can be characterized by some bounded (locally "regular") reference configuration, having the meaning of the "ground state", and its H^1 perturbations. Each island is therefore isomorphic to a Hilbert space (*Hilbert space sector*).

The stability under time evolution is guaranteed by the condition that the reference configuration satisfies a generalized stationarity condition, i.e.

it solves some elliptic problem. Such a condition is in particular satisfied by the time independent solutions and *a fortiori* by the minima $\overline{\varphi}$ of the potential whose corresponding Hilbert space sectors $\mathcal{H}_{\overline{\varphi}}$ are of the form $\{\overline{\varphi} + \chi, \chi \in H^1\}$. The existence of minima of the potential unstable under the symmetry therefore gives rise to phases or disjoint physical worlds in which the *symmetry* cannot be realized or, as one says, is *spontaneously broken*. This mechanism crucially involves both the asymmetry of the ground state *and* the infinite extension of the system, with no analog in the finite dimensional case.

This phenomenon is deeply rooted in the non-linearity of the problem and the fact that infinite degrees of freedom are involved. A simple prototype is given by the non-linear wave equation for a Klein-Gordon field $\varphi : \mathbf{R}^s \rightarrow \mathbf{R}^n$, with "potential" $U(\varphi) = \lambda(\varphi^2 - a^2)^2$. The model displays some analogy with the mechanical model of a particle in $\mathbf{R}^n$ subject to the potential $U(q) = \lambda(q^2 - a^2)^2$, which can be regarded as the higher dimensional version of the one-dimensional double well potential. But the differences are substantial: in the infinite dimensional case of the Klein-Gordon field, each point q has actually become infinite dimensional and, in fact, each absolute minimum $\overline{\varphi}$, with $|\overline{\varphi}| = a$ identifies the infinite set of configurations which have this point as asymptotic limit, namely the Hilbert space of configurations which are H^1 modifications of $\overline{\varphi}$. Whereas in the finite dimensional case there is no physical obstruction or "barrier", which prevents the motion from one minimum to the other, in the infinite dimensional case there is no physically realizable operation which leads from the Hilbert space sector defined by one minimum to that defined by another minimum, because this would require to change the asymptotic limit of the configurations and this is not possible by means of essentially localized operations, the only ones which are physically realizable. Pictorially, one could say that one cannot change the boundary conditions of the "universe" or of the (infinite volume) thermodynamical phase in which one is living.

The realization of the above structures allows to evade part of the conclusions of the standard textbook presentations of Noether's theorem and to account for spontaneous symmetry breaking; the point is that the standard presentations of the theorem do not consider the possibility of disjoint sectors unstable under the symmetry of the Hamiltonian and implicitly assume that the solutions vanish at infinity. In fact, one may prove that the *local conservation law*, $\partial^\mu j_\mu(x) = 0$, associated with a given symmetry of the Hamiltonian or of the Lagrangean, gives rise to a *global conservation law* or to a conserved "charge", which acts as the generator of the symmetry transformations for all the elements of a given Hilbert space sector $\mathcal{H}_{\overline{\varphi}}$, only if the symmetry leaves the sector invariant. Thus, only the stability subgroup of the given phase admits time independent generators in that phase, given by the charges of the corresponding Noether currents.

Clearly, if G is the (concrete) group of transformations which commutes with the time evolution, the whole set of solutions of the non-linear dynamical problem can be classified in terms of irreducible representations (or multiplets) of G, but if G is spontaneously broken in a given island defined by the Hilbert space sector $\mathcal{H}_{\overline{\varphi}}$, the latter cannot be the carrier of a representation of the symmetry group G, and in particular the elements of $\mathcal{H}_{\overline{\varphi}}$ cannot be classified in terms of multiplets of G.

One might think of grouping together solutions corresponding to initial data of the form $\overline{\varphi} + g\chi$, $g \in G$, which might look like candidates for multiplets of G. As a matter of fact, such sets of initial data do correspond to representations of a group of transformations which is isomorphic to G, but which does not commute with the dynamics, and therefore the above form of the initial data does not extend to arbitrary times; thus the above identification of multiplets at the initial time is not stable under time evolution. As a matter of fact, the group of transformations which commute with the time evolution corresponds to $\overline{\varphi} + \chi \rightarrow g\overline{\varphi} + g\chi$, $g \in G$, which, however, does not leave $\mathcal{H}_{\overline{\varphi}}$ stable.

Within this approach, it is possible to prove a classical counterpart of the so-called Goldstone theorem, according to which there are massless modes (i.e. solutions of the free wave equation) associated to each broken generator. The theorem proved here provides a mathematically acceptable substitute of the heuristic arguments and improves the conclusions based on the quadratic approximation of the potential around an absolute minimum.

Explicit examples which illustrate how these ideas work in concrete models are discussed in Chap. 8.

The discussion of symmetry breaking in classical systems relies, with some additions, on papers written jointly with Cesare Parenti and Giorgio Velo, to whom I am greatly indebted (see the references at the relevant points). An attempt is made to reduce the mathematical details to the minimum required to make the arguments self-contained and also convincing for a mathematically-minded reader. The required background technical knowledge is kept to a rather low level, in order that the lectures be accessible also to undergraduate students with a basic knowledge of Hilbert space structures.

1 Symmetries of a Classical System

The realization of symmetries in physical systems has proven to be of help in the description of physical phenomena: it makes it possible to relate the behaviour of similar systems and therefore it leads to a great simplification of the mathematical description of Nature.

The simplest concept of symmetry occurs at the geometrical or kinematical level when the shape of an object or the configuration of a physical system is invariant or symmetric under geometric transformations like rotations, reflections etc.. At the dynamical level, a system is symmetric under a transformation of the coordinates or of the parameters which identify its configurations, if correspondingly its dynamical behaviour is symmetric in the sense that the action of the symmetry transformation and of time evolution commute.

To formalize the concept of dynamical symmetry, we first recall that the *description of a classical physical system* consists in

i) the identification of all its possible configurations $\{S_\gamma\}$, with γ running over an index set of coordinates or parameters which identify the configuration S_γ

ii) the determination of their time evolution

$$\alpha^t : S_\gamma \to \alpha^t S_\gamma \equiv S_{\gamma(t)}. \tag{1.1}$$

A *symmetry g* of a physical system is a transformation of the coordinates (or of the parameters) γ, $g : \gamma \to g\gamma$, which

1) induces an invertible mapping of configurations

$$g : S_\gamma \to g S_\gamma \equiv S_{g\gamma} \tag{1.2}$$

2) does not change the dynamical behaviour[1], namely

$$\alpha^t g S_\gamma = \alpha^t S_{g\gamma} \equiv S_{(g\gamma)(t)} = S_{g\gamma(t)} = g\alpha^t S_\gamma. \tag{1.3}$$

---

[1] To simplify the discussion, here we do not consider the more general case in which the dynamics transform covariantly under g (like e.g. in the case of Lorentz transformations). For a general discussion of symmetries and of their relevance in physics, see R.M.F. Houtappel, H. Van Dam and E.P. Wigner, Rev. Mod. Phys. **37**, 595 (1965).

F. Strocchi: *Symmetries of a Classical System*, Lect. Notes Phys. **732**, 7–8 (2008)
DOI 10.1007/978-3-540-73593-9_1 © Springer-Verlag Berlin Heidelberg 2008

The above condition states that the symmetry transformation commutes with time evolution. For classical canonical systems, this amounts to the invariance of the Hamiltonian under the symmetry g (*symmetric Hamiltonian*).

The realization of a symmetry which relates (the configurations of) two seemingly different systems clearly leads to a unification of their description. In particular, the solution of the dynamical problem for one configuration automatically gives the solution for the symmetry related configuration (see (1.3)).

Example 1.1. double well potential. Consider a particle moving on a line, subject to a double well potential, i.e. described by the following Hamiltonian

$$H = \tfrac{1}{2}p^2 + \tfrac{1}{4}\lambda(q^2 - a^2)^2, \tag{1.4}$$

with q, p the canonical coordinates which label the configurations of the particle. The reflection $g : q \to -q, \ p \to -p$ leaves the Hamiltonian invariant and is a symmetry of the system; obviously, it maps solutions (of the Hamilton equations) into solutions.

Now, consider the two classes $\Gamma_\pm$ of solutions corresponding to initial conditions in the neighborhoods of the two absolute minima $q_0 = \pm a$, with $p_0 < \sqrt{\lambda}a^2/2$ respectively, and suppose that by some (artificial) *ansatz*, in the preparation of the initial configurations one cannot dispose of energies greater than $\lambda\,a^4/4$. This means that the two classes of solutions describe two disjoint realizations of the system, in the sense that *by fiat* no physically realizable operation allows to change a configuration from one class to the other. In this way, one gets a picture similar to the case of the thermodynamical phases, which are physically disjoint in the thermodynamical limit, but nevertheless described by the same Hamiltonian and related by a symmetry which is not implementable in each phase. Clearly, the existence of such a symmetry, even if devoid of physical operational meaning, provides a unified description of the two "phases".

For a particle moving on a plane, the analog of the double well potential defines a Hamiltonian which is invariant under rotations around the axis (through the origin) orthogonal to the plane and one has a continuous group of symmetries. There is a continuous family of absolute minima lying on the circle $|q_0|^2 = a^2$. Since such minima are not separated by any energy barrier, one cannot associate with them different systems by some artificial *ansatz* as above.

2 Spontaneous Symmetry Breaking

One of the most powerful ideas of modern theoretical physics is the mechanism of spontaneous symmetry breaking. It is at the basis of most of the recent achievements in the description of phase transitions in Statistical Mechanics as well as of collective phenomena in solid state physics. It has also made possible the unification of weak, electromagnetic and strong interactions in elementary particle physics. Philosophically, the idea is very deep and subtle (this is probably why its exploitation is a rather recent achievement) and the popular accounts do not fully do justice to it.

Roughly, spontaneous symmetry breaking is said to occur when a symmetry of the Hamiltonian, which governs the dynamics of a physical system, does not lead to a symmetric description of the physical properties of the system. At first sight, this may look almost paradoxical. From elementary courses on mechanical systems, one learns that the symmetries of a system are displayed by the symmetries of the Hamiltonian, which describes its time evolution; how can it then be that a symmetric Hamiltonian gives rise to an asymmetric physical description of a dynamical system?

The cheap standard explanation is that such a phenomenon is due to the existence of a non-symmetric absolute minimum or "ground state", but the mechanism must have a deeper explanation, since the symmetry of the Hamiltonian implies that an asymmetric stable point cannot occur by alone, (the action of the symmetry on it will produce another stable point). Now, the existence of a set of absolute minima related by a symmetry (or "degenerate ground states") does not imply a non-symmetric physical description. One actually gets a symmetric picture, if the correct correspondence is made between the configurations of the system (and their time evolutions), and such a correspondence is physically implementable if for any physically realizable configuration its transformed one is also realizable.

The way out of this argument is to envisage a mechanism by which, given a non-symmetric absolute minimum (or "ground" state) S_0, there are physical obstructions to reach its transformed one, $g \, S_0$, by means of physically realizable operations, so that effectively one gets confined to an asymmetric realization of the system. The purpose of the following discussion is to make such a rather vague and intuitive picture more precise.

F. Strocchi: *Spontaneous Symmetry Breaking*, Lect. Notes Phys. **732**, 9–11 (2008)
DOI 10.1007/978-3-540-73593-9_2   © Springer-Verlag Berlin Heidelberg 2008

For a classical finite dimensional dynamical system, two configurations may be said to be related by *physically realizable operations* if there is no physical obstruction for operationally changing one into the other, e.g. if they are connected by a continuous path of configurations, all with finite energy. In this way, one gets a partition of the configurations into classes and given a configuration S, the set of configurations which can be reached from it, by means of physically realizable operations, will be called the *phase Γ_S*, or the "physical world", to which S belongs.

A *symmetry g* will be said to be *physically realized* (or *implementable* or *unbroken*), in the phase Γ, if it leaves Γ stable.

In the mechanical example of the double well potential discussed above, there is no natural and physically reasonable way of isolating the solutions in the neighborhoods of the two minima, since an artificial limitation of the available energies looks rather unphysical. Actually, according to the above definitions, there is only one phase and the reflection symmetry is physically implementable or unbroken.

In order to further illustrate the above definitions, we consider a particle moving on a line, subject to a deformed double well potential, still invariant under the reflection $g : q \rightarrow -q$, with two absolute minima at $q_0 = \pm a$, but going to infinity as $q \rightarrow 0$.

Consider now two kinds of (one-dimensional) creatures, one living in the valley with bottom $q_0 = a$ and the other in the valley with bottom $q_0 = -a$. The infinite potential barrier prevents going from one valley to the other (tunneling is impossible); then, e.g., the people living in the r.h.s. valley do not have access to the l.h.s. valley, neither by action on the initial conditions of the particle nor by time evolution. Thus, the operations which are physically realizable (by each of the two kinds of people) cannot make transition from one valley to the other and the particle configurations get divided into two phases, labeled by the two minima Γ_a, Γ_{-a}, respectively.

The reflection symmetry is not physically realized in each of the two phases. As a matter of fact, even if the particle motion is described by a symmetric Hamiltonian, the particle physical world will look asymmetric to each kind of creatures: the *symmetry* is *spontaneously broken*.

The somewhat artificial example of spontaneous symmetry breaking discussed above is made possible by the infinite potential barrier between the two absolute minima. Clearly, such a mechanism is not available in the case of a continuous symmetry, since then the (absolute) minima are continuously related by the symmetry group and no potential barrier can occur between them (for a concrete example see the two dimensional double well discussed above). Thus, for finite dimensional classical dynamical systems, a continuous symmetry of the Hamiltonian is always unbroken (even if the ground state is degenerate and non-symmetric).

The often quoted example of a particle in a two dimensional double well potential is a somewhat misleading example of spontaneous breaking of continuous symmetry (it is also an incorrect example in one dimension, unless the potential is so deformed to produce an infinite barrier between the two minima). Actually, most of the claimed simple mechanical examples of spontaneous symmetry breaking discussed in the literature are equally misleading.

Even if the existence of non-symmetric minima is a rather peculiar phenomenon which deserves special interest, it does not imply spontaneous symmetry breaking in the radical sense of its realization in elementary particle physics, many body systems, statistical mechanics etc., where a symmetry of the dynamics is not shared by physical realizations or disjoint phases of the system. This is a much deeper phenomenon than the mere existence of *non-symmetric minima.*

The relevance of the distinction between *non-symmetric minima or ground states* and *spontaneous symmetry breaking* appears clear if one considers, e.g., a free particle on a line, where each configuration ($q_0 \in \mathbb{R}, p_0 = 0$) is a minimum of the Hamiltonian and it is not stable under translations, but nevertheless one does not speak of symmetry breaking; in fact, according to our definition, there is only one phase stable under translations.

The two concepts of symmetry breaking coincide for *infinitely extended systems*, since in this case, as we shall see below, different ground states define different phases or disjoint worlds; therefore their asymmetry necessarily leads to symmetry breaking in the radical sense of a non-symmetric physical description (see Chap. 7 below).

Similar considerations apply to classical systems which exhibit bifurcation[2] for which, strictly speaking, one does not have spontaneous symmetry breaking as long as the multiple solutions are related by physically realizable operations. As we shall see later, the latter property may fail if one considers the infinite volume (or thermodynamical) limit, and in this way spontaneous symmetry breaking may occur.

---

[2] D.H. Sattinger, Spontaneous Symmetry Breaking: mathematical methods, applications and problems in the physical sciences, in *Applications of Non-Linear Analysis*, H. Amann et al. eds., Pitman 1981.

3 Symmetries in Classical Field Theory

As the previous discussion indicates, it is impossible to realize the phenomenon of (spontaneous) breaking of a continuous symmetry in classical mechanical systems with a finite number of degrees of freedom. We are thus led to consider infinite dimensional systems, like classical fields.

Our main purpose is to recognize the existence of disjoint "phases", in the set of solutions of the classical field equations, with the interpretation of possible disjoint realizations of the system (Chapter 5). The phenomenon of spontaneous symmetry breaking in a given "phase" will then be explained by its instability under the symmetry transformation.

To simplify the discussion, we will focus our attention to the standard case of the non-linear equation

$$\Box \varphi + U'(\varphi) = 0, \tag{3.1}$$

where $\Box \equiv (\partial_t)^2 - \Delta$, $\varphi = \varphi(x, t)$, $x \in \mathbb{R}^s$, $t \in \mathbb{R}$, is a field taking values in $\mathbb{R}^n$, (an n-component field), $U(\varphi)$ is the potential, which for the moment will be assumed to be sufficiently regular, and U' denotes its derivative.

Equation (3.1) can be derived by the stationarity of the following action integral

$$\mathcal{A}(\varphi, \dot{\varphi}) = \int d^s x \, dt \, [-\tfrac{1}{2}(\nabla \varphi)^2 + \tfrac{1}{2}\dot{\varphi}^2 - U(\varphi)].$$

A typical prototype is given by

$$U(\varphi) = \tfrac{1}{4}\lambda(\varphi^2 - a^2)^2 \tag{3.2}$$

which is the infinite dimensional version of the double well potential discussed in Chap. 1.

Quite generally, (3.1) occurs in the description of non-linear waves in many branches of physics like non-linear optics, plasma physics, hydrodynamics, elementary particle physics etc.[3]. The above equation (3.1) will be used to illustrate general structures likely to be shared by a large class of non-linear hyperbolic equations.

---

[3] See, e.g., G.B. Whitham, *Linear and Non-Linear Waves*, J. Wiley, New York 1974; R. Rajaraman, Phys. Rep. **21 C**, 227 (1975); S. Coleman, *Aspects of Symmetry*, Cambridge Univ. Press 1985, Chap. 6

F. Strocchi: *Symmetries in Classical Field Theory*, Lect. Notes Phys. **732**, 13–16 (2008)
DOI 10.1007/978-3-540-73593-9_3   © Springer-Verlag Berlin Heidelberg 2008

The solution of the Cauchy problem for the (in general non-linear) equation (3.1), with given initial data

$$\varphi(x, t = 0) = \varphi_0(x), \qquad \partial_t \varphi(x, t = 0) = \psi_0(x), \tag{3.3}$$

provides the corresponding classical field $\varphi(x, t)$ described by (3.1).

In analogy with the previous discussion of the finite dimensional systems, a description of the system (3.1) consists in the identification of the class of initial conditions, for which the time evolution is well defined. Deferring the mathematical details, we will now denote by X the functional space within which the Cauchy problem is well posed, i.e. such that for any initial data

$$u_0 = \begin{pmatrix} \varphi_0 \\ \psi_0 \end{pmatrix} \in X \tag{3.4}$$

there is a unique solution $u(x, t)$ continuous in time (in the topology of X, see below) and belonging to X for any t, briefly $u(x, t) \in C^0(X, \mathbb{R})$.

Thus, X can be regarded as describing the initial configurations of the system (3.1) and it is stable under time evolution. [4]

In analogy with the finite dimensional case, a *symmetry* of the system (3.1) is an invertible mapping T_g of X onto X, which commutes with the time evolution. To simplify the discussion, we will make the technical assumption that T_g is a continuous mapping (in the X topology) of the form

$$T_g \begin{pmatrix} \varphi(x) \\ \psi(x) \end{pmatrix} = \begin{pmatrix} g(\varphi(x)) \\ J_g(\varphi(x))\psi(x) \end{pmatrix}, \tag{3.5}$$

with g a diffeomorphism of $\mathbb{R}^n$ of class C^2 and J_φ the Jacobian matrix of g. Such symmetries are called *internal symmetries*, since they commute with space and time translations.[5]

Under general regularity assumptions on the potential, such that for infinitely differentiable initial data the corresponding solution of (3.1) is of class C^2 in the variables x and t, one gets a characterization of the internal symmetries of the system (3.1).

---

[4] For an extensive review on the mathematical problems of the non-linear wave equation see M. Reed, *Abstract non-linear wave equation*, Springer-Verlag, Heidelberg 1976. For the solution of the Cauchy problem for initial data not vanishing at infinity, a crucial ingredient for discussing spontaneous symmetry breaking, see C. Parenti, F. Strocchi and G. Velo, Phys. Lett. **59B**, 157 (1975); Ann. Scuola Norm. Sup. (Pisa), III, 443 (1976), hereafter referred as I. A simple account with some addition is given in F. Strocchi, in *Topics in Functional Analysis 1980-81*, Scuola Normale Superiore Pisa, 1982. For a beautiful review of the recent developments see W. Strauss, *Nonlinear Wave Equations*, Am. Math. Soc. 1989.

[5] For the discussion of more general symmetries see C. Parenti, F. Strocchi and G. Velo, Comm. Math. Phys. **53**, 65 (1977), hereafter referred as II; Phys. Lett. **62B**, 83 (1976).

Theorem 3.1. [6] *Under the above assumption on U, any internal symmetry of the system (3.1) is characterized by a g which is an affine transformation*

$$g(z) = Az + a, \tag{3.6}$$

where $a \in \mathbb{R}^n$ and A is an $n \times n$ invertible matrix. Furthermore, the invariance of the action integral up to a scale factor requires

$$A^T A = \lambda \mathbf{1}, \tag{3.7}$$

with A^T the transpose of A and λ a suitable constant. A, a, λ, which depend on g, satisfy the following condition,

$$U(Az + a) = \lambda U(z) + U(a). \tag{3.8}$$

Proof. The condition that $T_g \alpha^t u_0 = \alpha^t T_g u_0$ be a solution of (3.1), for any initial data u_0, implies[7]

$$0 = \Box g_k(\varphi) + U'_k(g(\varphi)) =$$

$$= \frac{\partial^2 g_k}{\partial z_i \partial z_j}(\varphi) \, \partial^\mu \varphi_i \partial_\mu \varphi_j - \frac{\partial g_k}{\partial z_i}(\varphi) U'_i(\varphi) + U'_k(g(\varphi)). \tag{3.9}$$

Choosing the initial data such that $\varphi_0(x) = \text{const} \equiv c$, $\psi_0(x) = 0$, for x in some region of $\mathbb{R}^s$, the first term of (3.9) vanishes there and one gets

$$-\frac{\partial g_k}{\partial z_i}(c) U'_i(c) + U'_k(g(c)) = 0. \tag{3.10}$$

Since c is arbitrary, the sum of the last two terms vanishes for any φ. Choosing now $\varphi_0(x) = c$, $\psi_0(x) = \text{const} = b$, $x \in V \subset \mathbb{R}^s$, one gets

$$\frac{\partial^2 g_k}{\partial z_i \partial z_j}(c) = 0, \quad \forall c \in \mathbb{R}^n, \quad \text{i.e. } g(z) = Az + a.$$

Equation (3.9) then becomes

$$\frac{\partial}{\partial z_l} U(Az + a) = (A^T A)_{li} \frac{\partial}{\partial z_i} U(z).$$

The invariance of the action integral up to a scale factor requires $A^T A = \lambda \mathbf{1}$ and $U(Az + a) = \lambda U(z) + \text{const}$; the normalization $U(0) = 0$ identifies the latter constant as $U(a)$.

Having characterized the possible symmetries of (3.1), we may now ask whether symmetry breaking can occur. For continuous groups this possibility seems to be in conflict with Noether's theorem.

---

[6] Ref. II (see footnote 5).

[7] We use the convention by which sum over dummy indices is understood; furthermore the relativistic notation is used: $\mu = 0, 1, 2, 3$, $\partial_0 = \partial/\partial t$, $\partial_i = \partial/\partial x^i$, $i = 1, 2, 3$, $\partial^\mu = g^{\mu\nu}\partial_\nu$, $g^{00} = 1 = -g^{ii}$, $g^{\mu\nu} = 0$ if $\mu \neq \nu$.

Theorem 3.2. [8] *Let G be an N parameter Lie group of internal symmetries for the classical system* (3.1), *then there exist N conserved currents*

$$\partial^\mu J_\mu^a(x,t) = 0, \qquad a = 1, ..., N \qquad (3.11)$$

and N conserved quantities

$$Q^a(t) = \int d^s x \, J_0^a(x,t) = Q^a(0), \qquad (3.12)$$

which are the generators of the corresponding one-parameter subgroups $\{g_\alpha^a, \, \alpha \in \mathbb{R}\}$ of symmetry transformations

$$\delta^a u \equiv dg_\alpha^a(u)/d\alpha|_{\alpha=0} = \{u, Q^a\}, \qquad (3.13)$$

where the curly brackets denote the Poisson brackets.

For the proof we refer to any standard textbook.[9] [10]

One should stress that for (3.12) some regularity properties of the solution are needed, even if they are not spelled out in the standard accounts of the theorem. Actually, the deep physical question of spontaneous breaking requires a more refined analysis of the mathematical properties of the solutions and of their behaviour at infinity. As we shall see, the problem of existence of "islands" or phases, stable under time evolution (playing the role of the valleys of the example discussed in Chap. 2) and characterized by a non-trivial behaviour at infinity of the corresponding solutions will require a sort of stability theory for the infinite dimensional system (3.1).

---

[8] E. Noether, Nachr. d. Kgl. Ges. d. Wiss. Göttingen (1918), p. 235.

[9] See e.g. H. Goldstein, *Classical Mechanics*, 2nd. ed., Addison-Wesley 1980; E. L. Hill, Rev. Mod. Phys. **23**, 253 (1951); N.N. Bogoljubov and D.V. Shirkov, *Introduction to the theory of quantized fields*, Interscience 1958, Sect. 2.5.

[10] For the representations of Lie groups and their generators in classical systems, see D.G. Currie, T.F. Jordan and E.C.G. Sudarshan, Rev. Mod. Phys. **35**, 350 (1963); E.C.G. Sudarshan and N. Mukunda, *Classical Dynamics: A Modern Perspective*, J. Wiley and Sons 1974.

4 General Properties of Solutions of Classical Field Equations

The first basic question is to identify the possible configurations of the systems (3.1), namely the set X of initial data for which the time evolution is well defined and which is mapped onto itself by time evolution. In the mathematical language, one has to find the functional space X for which the Cauchy problem is well posed. In order to see this, one has to give conditions on $U'(\varphi)$ and to specify the class of initial data or, equivalently, the class of solutions one is interested in. Here one faces an apparently technical mathematical problem, which has also deep physical connections.

In the pioneering work by Jörgens[11] and Segal[12], the choice was made of considering those initial data (and, consequently, those solutions) for which the total "kinetic" energy is finite[13]

$$E_{kin} \equiv \tfrac{1}{2} \int [(\nabla\varphi)^2 + \varphi^2 + \psi^2] \, d^s x < \infty, \quad \psi = \dot{\varphi}. \qquad (4.1)$$

From a physical point of view, condition (4.1) is unjustified and it automatically rules out very interesting cases, like the external field problem, the symmetry breaking solutions, the soliton-like solutions and, in general, all the solutions which do not decrease sufficiently fast at large distances to make the above integral (4.1) convergent. Actually, there is no physical reason why E_{kin} should be finite, since even the splitting of energy into a kinetic and a potential part is not free of ambiguities. Therefore, we have to abandon condition (4.1) and we only require that the initial data are *locally* smooth

---

[11] K. Jörgens, Mat. Zeit. **77**, 291 (1961).

[12] I. Segal, Ann. Math. **78**, 339 (1963).

[13] Strictly speaking, the kinetic energy should not involve the term φ^2. Our abuse of language is based on the fact that the bilinear part of the total energy corresponds to what is usually called the "non-interacting" theory (whose treatment is generally considered as trivial or under control by an analysis in terms of normal modes). The remaining term in the total energy is usually considered as the true interaction potential.

F. Strocchi: *General Properties of Solutions of Classical Field Equations*, Lect. Notes Phys.
732, 17–20 (2008)
DOI 10.1007/978-3-540-73593-9_4

in the sense that

$$\int_V [(\nabla \varphi)^2 + \varphi^2 + \psi^2] \, d^s x < \infty \qquad (4.2)$$

for any bounded region V (*locally finite kinetic energy*).

As it is usual in the theory of second order differential equations, one may write (3.1) in first order (or Hamiltonian) formalism, by grouping together the field $\varphi(t)$ and its time derivative $\psi(t) = \dot{\varphi}(t)$ in a two component vector

$$u(t) = \begin{pmatrix} \varphi(t) \\ \psi(t) \end{pmatrix} \equiv \begin{pmatrix} u_1(t) \\ u_2(t) \end{pmatrix}.$$

Equation (3.1) can then be written in the form

$$\frac{du}{dt} = Ku + f(u), \qquad (4.3)$$

with the initial condition

$$u(0) = u_0 = \begin{pmatrix} \varphi_0 \\ \psi_0 \end{pmatrix}, \qquad (4.4)$$

where

$$K = \begin{pmatrix} 0 & 1 \\ \triangle & 0 \end{pmatrix}, \quad f(u) = \begin{pmatrix} 0 \\ -U'(\varphi) \end{pmatrix}. \qquad (4.5)$$

One of the two components of (4.3) is actually the statement that $\psi = \dot{\varphi}$.

It is more convenient to rewrite (4.3) as an integral equation which incorporates the initial conditions. To this purpose, we introduce the one parameter continuous group $W(t)$ generated by K and corresponding to the free wave equation (see Appendix A)

$$W(0) = 1, \qquad W(t + s) = W(t) \, W(s) \qquad \forall t, s.$$

Then, the integral form of (4.3) is

$$u(t) = W(t)u_0 + \int_0^t W(t - s)f(u(s))ds. \qquad (4.6)$$

The main advantage of (4.6) is that, in contrast to (4.3), it does not involve derivatives of u and, as we will see, it is easier to give it a precise meaning.

In first order formalism, the condition that the kinetic energy is locally finite reads: $u_1 = \varphi \in H^1_{loc}(\mathbb{R}^s)$, (i.e. $|\nabla \varphi|^2 + |\varphi|^2$ is a locally integrable function); $u_2 = \psi \in L^2_{loc}(\mathbb{R}^s)$. Thus, we assume the following *local regularity condition of the initial data*

$$u \in H^1_{loc}(\mathbb{R}^s) \oplus L^2_{loc}(\mathbb{R}^s) \equiv X_{loc}. \qquad (4.7)$$

The space X_{loc} is equipped with the natural topology generated by the family of *Hilbert seminorms*

$$\|u\|_V^2 = \int_V ((\nabla\varphi)^2 + \varphi^2)d^sx + \int_V \psi^2 d^sx. \tag{4.8}$$

Thus, X_{loc} is the Fréchet space defined as the inductive limit of the Hilbert spaces $\mathcal{H}(V)$ with Hilbert products (4.8).

As in the finite dimensional case, in order to solve the Cauchy problem we need some kind of Lipschitz condition[14] on the potential; in agreement with the local structure discussed above, it is natural to chose the following local condition.

Local Lipschitz Condition

a) $f(u)$ defines a continuous mapping of X_{loc} into X_{loc}
b) for any sphere Ω_R, of radius R, and for any $\rho > 0$, there exists a constant $C(\Omega_R, \rho)$, such that

$$\|f(u_1) - f(u_2)\|_{\Omega_R} \leq C(\Omega_R, \rho) \|u_1 - u_2\|_{\Omega_R}, \tag{4.9}$$

for all $u_1, u_2 \in X_{loc}$ such that $\|u_i\|_{\Omega_R} \leq \rho, i = 1, 2$ and

$$\sup_{0 \leq t \leq R/2} C(\Omega_{R-t}, \rho) \equiv \bar{C}(\Omega_R, \rho) < \infty.$$

The above local Lipschitz condition is satisfied by a large class of potentials U:

i)   in $s = 1$ dimension, if $U(\varphi)$ is an entire function;
ii) for $s = 2$, if

$$U(\varphi) = \sum_{\alpha \in \mathbb{N}^n}^{\infty} C_\alpha \varphi^\alpha, \tag{4.10}$$

α being a multi-index, $\varphi^\alpha = \varphi_1^{\alpha_1}...\varphi_n^{\alpha_n}$, with

$$\sum_{\alpha \in \mathbb{N}^n} |C_\alpha| \, |\alpha|^{|\alpha|/2} |\varphi|^{|\alpha|} < \infty,$$

iii) for $s = 3$, if U is a twice differentiable real function such that

$$\sup_\varphi (1 + |\varphi|^2)^{-1} |U''(\varphi)| < \infty. \tag{4.11}$$

The proof that the above classes of potentials satisfy the local Lipschitz condition is similar to that for global Lipschitz continuity (see Lemma 5.3 in Chap. 5), except that local Sobolev inequalities are used instead of global ones (for details see Ref. I, quoted in Chap. 3).

Since, for the present purposes, we are not interested in optimal conditions, (for a more general discussion see Ref. I), in the following discussion, for simplicity, we will consider potentials belonging to the above classes, for $s = 1, 2, 3$.

---

[14] See e.g. V. Arnold, *Ordinary Differential Equations*, Springer 1992, Chap. 4; G. Sansone and R. Conti, *Non-linear Differential Equations*, Pergamon Press 1964.

The above Local Lipschitz condition guarantees that

1) (4.6) is well defined for $u \in C^0(X_{loc}, \mathbb{R})$
2) the *solution* of (4.6), if it exists, *is unique*
3) (4.6) has an *hyperbolic character*, i.e. the local norm of $u(t)$ in the sphere Ω_{R-t} of radius $R - t$, $0 < t < R$, depends only on the local norm of $u(0)$ in the sphere Ω_R of radius R (the influence domain)

$$\|u(t)\|_{\Omega_{R-t}} \le Ae^{\omega t}\|u(0)\|_{\Omega_R}, \tag{4.12}$$

(ω a suitable constant)

4) *solutions* of (4.6) *exist for sufficiently small times.*

For the proof of 1) – 4), see Appendix B.

To continue the solutions from small times to all times, and in this way get a *global in time solution of the Cauchy problem*, one needs a bound which implies that the norm of $u(t)$ stays finite. This is guaranteed if U satisfies the following

Lower Bound Condition There exist suitable non-negative constants α, β such that

$$U(\varphi) \ge -\alpha - \beta|\varphi|^2. \tag{4.13}$$

In conclusion we have

Theorem 4.1. *(Cauchy problem: global existence of solutions)*[15]. *If U is such that the local Lipschitz condition and the lower bound condition are satisfied, then (4.6) has one and only one solution $u(t) \in C^0(X_{loc}, \mathbb{R})$.*

For a brief sketch of the proof see Appendix C.

---

[15] To our knowledge the proof of global existence of solutions of (4.6) for initial data in $H^1_{loc} \oplus L^2_{loc}$ first appeared in Ref. I, although the validity of such a result was conjectured by W. Strauss, Anais Acad. Brasil. Ciencias **42**, 645 (1970), p. 649, Remark: "The support restrictions on $u_0(x), u_1(x), F(x, t, 0)$ could probably be removed by exploiting the hyperbolic character of the differential equation . . . ".

5 Stable Structures, Hilbert Sectors, Phases

The mathematical investigation of the existence of solutions for the non-linear (4.6) does not exhaust the problem of the physical interpretation of the corresponding classical field theory. For infinitely extended systems, in general not every solution is physically acceptable; one has to supplement the analysis of the possible solutions by a list of mathematical properties which the solutions must share in order to allow a physical interpretation.

For quantum field theory the realization of the basic mathematical structure which renders the theory physically sound is due to Wightman[16] and it is nowadays standard to accept as "solutions" of the quantum field equations those which satisfy Wightman's axioms. A similar problem arises in Statistical Mechanics and the basic structure has been clarified[17], with the realization that the same dynamics may describe different physical realizations or phases of a given system.

Since not all solutions of (4.6) describe physically acceptable configurations, one has to look for those subsets S which satisfy a few (additional) basic requirements. In particular, we are interested in characterizing possible disjoint subsets S (of solutions), which may be interpreted as describing *disjoint realizations or "phases" of the system*, in strict analogy with the disjoint inequivalent representations of the algebra of local observables in Statistical Mechanics and in the Quantum Theory of infinitely extended systems.

With these motivations, general considerations, to be further discussed below, suggest to look for structures S, in the set of solutions of (4.6), characterized by the following properties

I   (**Local structure**) For a reasonable interpretation of S as a "phase" or a physical realization of the system, any two elements of S should be related by "physically realizable operations", in the sense, to be made more precise below, that they should describe configurations of the system both realizable or accessible in the same "laboratory" or phase. This implies that the two solutions can only differ locally, but not by their

---

[16] R.F. Streater and A.S. Wightman, *PCT, Spin and Statistics and All That,* Benjamin-Cumming Publ. C. 1980.

[17] See e.g. D. Ruelle, *Statistical Mechanics,* Benjamin 1969; R. Haag, *Local Quantum Theory,* Springer-Verlag 1992.

F. Strocchi: *Stable Structures, Hilbert Sectors, Phases,* Lect. Notes Phys. **732**, 21–28 (2008)
DOI 10.1007/978-3-540-73593-9_5   © Springer-Verlag Berlin Heidelberg 2008

behaviour at infinity, since by physically realizable operations one cannot change the boundary conditions of the "universe" or of the (infinite volume) thermodynamical phase in which one is living.

II **Stability of S under time evolution.**

III **Stability of S under space translations.**

One may also require that the infinite volume integral of the (renormalized) energy-momentum density is finite for each element of S. Of particular physical interest are those sets S which satisfy the following further condition

IV **Energy bounded from below** in S.

To be more precise we have to give a mathematical formalization of the above requirements.

I. **Local structure.** In order to convert condition I into a mathematical statement, one must formalize the intuitive idea of physically realizable operations. Since our measuring apparatuses and our possible operations on a physical system extend over *bounded* regions of space, starting from a given field configuration u, by physically realizable operations we can modify it only locally, i.e. we can reach only those configurations which essentially differ from u only locally (*quasi local modifications*). From a mathematical point of view, it is natural to identify the concept of quasi local modification as a $H^1(\mathbb{R}^s) \oplus L^2(\mathbb{R}^s)$ perturbation, i.e. given a solution $u_1(t)$, a solution $u_2(t)$ is a quasi local modification of u_1 if $u_1(t) - u_2(t) \in H^1(\mathbb{R}^s) \oplus L^2(\mathbb{R}^s)$ continuously in t, briefly

$$u_1(t) - u_2(t) \in C^0(H^1(\mathbb{R}^s) \oplus L^2(\mathbb{R}^s), \mathbb{R}). \tag{5.1}$$

We are thus led to introduce the following

Definition 5.1. *Let $\mathcal{F}$ denote the family of solutions $u(t) \in C^0(X_{loc}, \mathbb{R})$ of (4.6), a subset $S \subset \mathcal{F}$ has a* **local structure** *if (5.1) holds $\forall u_1, u_2 \in S$.*

As it appears also in other fields, the concept of "locality" plays an important rôle for the infinite dimensional generalization of ideas developed for finite dimensional systems. The emphasis on local structures is actually the key which makes possible (and physically meaningful) the treatment of the dynamics of infinite degrees of freedom. A crucial property is the stability of the local structure under time evolution.

II. **Stability under time evolution.** Since time evolution is one of the possible realizable "operations", the above definition of local structure is physically meaningful provided it is stable under time evolution, namely if $\forall u(t) \in S$ also $u_\tau(t) \equiv u(t + \tau) \in S$, $\forall \tau \in \mathbb{R}$. A set S with local structure satisfying such stability under time evolution will be called a **sector**. Thus, all the elements u of the sector S identified by a reference element $\bar{u}$ have the property that $\delta(t) \equiv u(t) - \bar{u}(0) \in H^1(\mathbb{R}^s) \oplus L^2(\mathbb{R}^s)$, $\forall t \in \mathbb{R}$.

Every solution $u(t) \in \mathcal{F}$ defines a set with a local structure (at worst that consisting of just one element), but in general it does not define a sector. In the latter case, the time evolution has a somewhat catastrophic character, since it drastically changes the large distance behaviour of the initial data; as we will discuss below, this would mean a change from one "phase" or physical world to another and this makes a reasonable physical interpretation difficult.

In general a sector S does not have a linear structure, nor that of the affine space $\bar{u}(0) + H^1 \oplus L^2$, since it is not guaranteed that for all $\delta_0 \in H^1 \oplus L^2$, the solution $u(t)$ corresponding to the initial data $\bar{u}(0) + \delta_0$ will belong to S. A sector with such a property is isomorphic to a Hilbert space and it is called a **Hilbert space sector** (HSS).

The existence of Hilbert space sectors is therefore controlled by the following stability problem: if two configurations $u_1(0)$, $u_2(0)$ are "close" at $t = 0$, in the sense that they differ by a quasi local perturbation, namely $u_1(0) - u_2(0) \in H^1(\mathbb{R}^s) \oplus L^2(\mathbb{R}^s)$, under which conditions will they remain "close" at any later times (and, therefore, are elements of a sector)?[18]

We defer the discussion of conditions III and IV to the next section. Now, we discuss the mathematical characterization of Hilbert space sectors and the conditions that guarantee their existence. The obvious questions are:

i) given a non-linear equation (4.6), can one *a priori* characterize the existence of non-trivial sectors associated to it? In particular, without having to solve (4.6), under which conditions (if any) can a set of initial data define a sector and what is its explicit content?

ii) can one characterize the existence and the structure of Hilbert space sectors, in the set of solutions of (4.6)?

One of the main conclusions of the analysis of this and the following Chapter is

Proposition *The constant solutions* $u(t) = u_0 = (\varphi_0, \; 0)$, *corresponding to the absolute minima* φ_0 *of the potential, define Hilbert space sectors* $\mathcal{H}_{\varphi_0}$ *consisting of the solutions which correspond to the set of initial data* $u_0 + H^1(\mathbb{R}^s) \oplus L^2(\mathbb{R}^s)$. *Furthermore, in each such a HSS all the solutions* $\varphi(\mathbf{x}, t)$ *have the same asymptotic limit* φ_0, *for* $|\mathbf{x}| \to \infty$ *and the energy is bounded from below.*

Such a characterization clarifies the basic difference between the finite and the infinite dimensional cases. Whereas in the first case the minima of the potential describe configurations with no physical obstruction or "barrier" which prevents a motion from one to the other, in the latter case each minimum identifies a Hilbert space of solutions, which is stable under time evolution and it is physically disjoint from the others, as the thermodynamical phases, since no physically realizable operation can change the boundary condition of the "universe", i.e. the asymptotic behaviour of the solutions.

---

[18] It is not difficult to recognize the analogies with the stability theory, which plays a crucial role in the theory of non-linear phenomena, in the finite dimensional case; see e.g. G. Sansone and R. Conti, *Non-Linear Differential Equations*, Pergamon Press 1964, Chap. IX.

For simplicity, we discuss the case in which the potential $U(\varphi)$ belongs to the following classes: it is an entire function in dimension $s = 1$ and it belongs to the classes (4.10) and (4.11) in dimension $s = 2, 3$, respectively. For a more general discussion see Ref. II.[19] Then we have

Theorem 5.2. *An initial data*

$$u_0 = \begin{pmatrix} \varphi_0 \\ \psi_0 \end{pmatrix} \in H^1_{loc} \oplus L^2_{loc}.$$

with φ_0 bounded, defines a non-trivial sector $\mathcal{H}_{u_0}$ iff

a) $\psi_0 \in L^2(\mathbb{R}^s),$ (5.2)

b) $\Delta\varphi_0 - U'(\varphi_0) \equiv h \in H^{-1}(\mathbb{R}^s),$ (5.3)

(i.e. the Fourier transform $\tilde{h}(k)$ satisfies $\int |\tilde{h}(k)|^2(1 + k^2)^{-1}d^s k < \infty$).

Actually, $\mathcal{H}_{u_0}$ is completely specified as the set of all solutions $v(t)$ with initial data of the form

$$v_0 = \begin{pmatrix} \varphi_0 + \chi \\ \psi_0 + \zeta \end{pmatrix}, \begin{pmatrix} \chi \\ \zeta \end{pmatrix} \in H^1(\mathbb{R}^s) \oplus L^2(\mathbb{R}^s),$$ (5.4)

*i.e. $\mathcal{H}_{u_0}$ is the affine space $u_0 + H^1(\mathbb{R}^s) \oplus L^2(\mathbb{R}^s)$ and, being isomorphic to $H^1(\mathbb{R}^s) \oplus L^2(\mathbb{R}^s)$, carries a Hilbert space structure (**Hilbert space sector**).*

Proof. Let $v(t)$ be a solution $\in \mathcal{F}$ and $u_0 \equiv \begin{pmatrix} \varphi_0 \\ \psi_0 \end{pmatrix}$, then

$$\delta(t) = \begin{pmatrix} \chi(t) \\ \zeta(t) \end{pmatrix} \equiv v(t) - u_0$$ (5.5)

satisfies the following integral equation

$$\delta(t) = W(t)\delta_0 + L(t) + \int_0^t ds\, W(t - s)\, g(\delta(s)),$$ (5.6)

where

$$L(t) = (W(t) - 1)u_0 + \int_0^t ds\, W(t - s) \begin{pmatrix} 0 \\ -U'(\varphi_0) \end{pmatrix}$$ (5.7)

$$= \begin{pmatrix} \frac{1-\cos\sqrt{-\Delta}\,t}{-\Delta} & \frac{\sin\sqrt{-\Delta}\,t}{\sqrt{-\Delta}} \\ \frac{\sin\sqrt{-\Delta}\,t}{\sqrt{-\Delta}} & \cos\sqrt{-\Delta}\,t - 1 \end{pmatrix} \begin{pmatrix} \Delta\varphi_0 - U'(\varphi_0) \\ \psi_0 \end{pmatrix} \equiv \begin{pmatrix} L_1(t) \\ L_2(t) \end{pmatrix},$$

$$g(\delta(s)) \equiv \begin{pmatrix} 0 \\ -G'_{\varphi_0}(\chi(s)) \end{pmatrix},$$ (5.8)

$$G_{\varphi_0}(\chi) \equiv U(\varphi_0 + \chi) - U(\varphi_0) - U'(\varphi_0)\chi.$$ (5.9)

---

[19] C. Parenti, F. Strocchi and G. Velo, Phys. Lett. **62B**, 83 (1976); Comm. Math. Phys. **53**, 65 (1977); Lectures at the Int. School of Math. Phys. Erice 1977, in *Invariant Wave Equations*, G. Velo and A.S. Wightman eds., Springer-Verlag 1978.

The subscript φ_0 and the explicit dependence on x through φ_0 will often be omitted in the sequel, using for brevity the notation $G(x, \chi(x))$ or simply $G(\chi)$. Furthermore, for brevity $\nabla_z G(x, z)\big|_{z=\chi}$ will be denoted by $G'(\chi)$.

The crux of the argument is that for φ_0 bounded, briefly $\in L^\infty(\mathbb{R}^s)$, for the class of potentials under consideration, $G(\chi)$ satisfies

i) $G'(\chi)$ is *globally Lipschitz continuous*, i.e. for any $\rho > 0$, there exists a constant $C(\rho)$ such that for any $\chi_1, \chi_2 \in H^1(\mathbb{R}^s)$, with $\|\chi_i\|_{H^1} \leq \rho, i = 1, 2$,

$$\|G'(\chi_2) - G'(\chi_1)\|_{L^2} \leq C(\rho)\|\chi_2 - \chi_1\|_{H^1} \tag{5.10}$$

ii) G satisfies a *lower bound condition*, i.e. there exists a non-negative constant γ, such that

$$G(x, z) \geq -\gamma|z|^2, \quad \forall z \in \mathbb{R}^n, x \in \mathbb{R}^s \tag{5.11}$$

(The proof of i) and ii) is given in Lemma 5.3 and 5.4, respectively).

Now, if i), ii) hold, since $g(0) = 0$, property i) implies that $g(\chi) \in H^1(\mathbb{R}^s) \oplus L^2(\mathbb{R}^s)$ and therefore, since $W(t)$ maps $H^1(\mathbb{R}^s) \oplus L^2(\mathbb{R}^s)$ into itself continuously in t, (see Appendix A),

$$\delta(t) \in C^0(H^1 \oplus L^2, \mathbb{R}) \quad \text{iff} \quad L(t) \in C^0(H^1 \oplus L^2, \mathbb{R}). \tag{5.12}$$

The latter condition is equivalent to conditions a) and b), ((5.2), (5.3)), (see Lemma 5.3 below).

The proof that the sector is not empty and actually is a Hilbert space sector amounts to proving that (5.6) has one and only one solution $\delta(t) \in C^\circ(H^1 \oplus L^2, \mathbb{R})$ for any initial data $\delta_0 \in H^1(\mathbb{R}^s) \oplus L^2(\mathbb{R}^s)$.

A simple important case is when u_0 is a static solution of (4.6),

$$\Delta\varphi_0 - U'(\varphi_0) = 0, \qquad \psi_0 = 0. \tag{5.13}$$

In this case $L(t) = 0$ and (5.6) has the same form of (4.6), for which the Cauchy problem in $H^1 \oplus L^2$ has been solved by Segal[20].

In the general case $L(t) \neq 0$, a generalization of Segal theorem (see Appendix D) gives existence and uniqueness in $H^1 \oplus L^2$.

Lemma 5.3. *For any $\varphi_0 \in L^\infty(\mathbb{R}^s)$, the function $G'(\chi)$ defined through (5.9) is globally Lipschitz continuous, (5.10).*

Proof. From the identity

$$G'(\chi^{(2)}) - G'(\chi^{(1)}) = U'(\varphi_0 + \chi^{(2)}) - U'(\varphi + \chi^{(1)})$$

$$= \int_0^1 d\sigma \frac{d}{d\sigma} U'(\varphi_0 + \chi^{(2)} + \sigma(\chi^{(2)} - \chi^{(1)}))$$

$$= \int_0^1 d\sigma \, U''(\varphi_0 + \chi^{(2)} + \sigma(\chi^{(2)} - \chi^{(1)}))(\chi^{(2)} - \chi^{(1)}),$$

---

[20] See footnote 12.

(5.10) will follow if, for any $\rho > 0$, there exists a constant $C(\rho)$ such that

$$\sup_{k=1,\ldots n} \|\sum_{j=1}^{n} \frac{\partial^2 U}{\partial z_j \partial z_k}(\varphi_0 + \chi')\chi_j\|_{L^2} \le C(\rho)\|\chi\|_{H^1}, \tag{5.14}$$

for all $\chi', \chi \in H^1$ with $\|\chi'\| \le \rho, \|\chi\| \le \rho$.

For the class of potentials under consideration, the proof of (5.14) reduces to the estimate of terms of the type $(\varphi + \chi^{(1)})^\alpha \chi^{(2)}$ with $\chi^{(i)} \in H^1, i = 1, 2$, $\alpha \in \mathbb{N}^n$ for $s = 1, 2$ and $|\alpha| \le 2$ for $s = 3$. Now, since $|a + b|^p \le 2^p(|a|^p + |b|^p)$, $\forall a, b \in \mathbb{R}$, $p \ge 1$, one has

$$\|(\varphi_0 + \chi^{(1)})^\alpha \chi^{(2)}\|_{L^2} \le 2^{|\alpha|}\{\||\varphi_0|^{|\alpha|} |\chi^{(2)}|\|_{L^2} + \||\chi^{(1)}|^{|\alpha|} |\chi^{(2)}|\|_{L^2}\} \tag{5.15}$$

and the first term on the r.h.s. is immediately estimated by

$$2^{|\alpha|}\||\varphi_0|^{|\alpha|} |\chi^{(2)}|\|_{L^2} \le A^{|\alpha|}(\|\varphi_0\|_{L^\infty})^{|\alpha|} \|\chi^{(2)}\|_{H^1}. \tag{5.16}$$

The second term can be estimated by using the usual Hölder and the Sobolev inequalities[21]

$$2^{|\alpha|}\||\chi^{(1)}|^{|\alpha|}|\chi^{(2)}|\|_{L^2} \le 2^{|\alpha|}\||\chi^{(1)}|\|_{L^{2(|\alpha|+1)}}^{|\alpha|}\||\chi^{(2)}|\|_{L^{2(|\alpha|+1)}}$$
$$\le B^{|\alpha|}C_s(2|\alpha| + 2)^{|\alpha|+1}\|\chi^{(1)}\|_{H^1}^{|\alpha|}\|\chi^{(2)}\|_{H^1}. \tag{5.17}$$

Thus for $s = 3$ the proof is completed. For $s = 1, 2$ the convergence of the sum over α is guaranteed by the properties which characterize the class of potentials under consideration.

Lemma 5.4. For $\varphi_0 \in L^\infty(\mathbb{R}^s)$, the lower bound condition for the potential, (4.13), implies that (5.11) holds.

Proof. Consider the identity

$$G(y) = \int_0^1 d\sigma(1 - \sigma)\frac{d^2}{d\sigma^2}U(\varphi_0 + \sigma y) = \int_0^1 d\sigma(1 - \sigma)y^2 U''(\varphi_0 + \sigma y). \tag{5.18}$$

---

[21] See e.g. L.R. Volevic and B.P. Paneyakh, Russian Math. Surveys **20**, 1 (1965). We list them for the convenience of the reader

$$s = 1, \quad \|f; L^p(\mathbb{R}^1)\| \le C_1(p)\|f; H^1(\mathbb{R}^1)\|, \quad 2 \le p \le \infty, C_1(p) = 0(1),$$
$$s = 2, \quad \|f; L^p(\mathbb{R}^2)\| \le C_2(p)\|f; H^1(\mathbb{R}^2)\|, \quad 2 \le p < \infty, C_2(p) = 0(p^{\frac{1}{2}}),$$
$$s = 3, \quad \|f; L^p(\mathbb{R}^3)\| \le C_3(p)\|f; H^1(\mathbb{R}^3)\|, \quad 2 \le p \le 6, \quad C_3(p) = 0(1).$$

The same kind of estimates hold locally. In particular, for any cube $K \subset \mathbb{R}^s$ of size R, they take the form

$$\|f; L^p(K)\| \le C_{s,R}(p)\|f; H^1(K)\|,$$

with $p \in [2, +\infty]$ for $s = 1$, $p \in [2, +\infty[$ for $s = 2$ and $p \in [2, 6]$ for $s = 3$. The constants $C_{s,R}(p)$ depend only on the size R and exhibit the same dependence on p as in the global case.

Since U is of class C^2, and φ_0 is bounded, $U''(\varphi_0 + \sigma y)$ is bounded below for $|y| \leq 1$, $0 \leq \sigma \leq 1$; hence from (5.14) we get a lower bound for G of the form of (5.11). On the other hand, for $|y| \geq 1$, the lower bound condition (4.13), gives

$$G(y) \geq - \{\alpha + \beta + \beta \sup_{x \in \mathbb{R}^s} [|\varphi_0(x)|^2 + 2|\varphi_0(x)| + U'(\varphi_0(x))]$$

$$+ \max(0, \sup_{x \in \mathbb{R}^s} U(\varphi_0(x)))\}|y|^2$$

Lemma 5.5. $L(t) \in C^0(H^1 \oplus L^2, \mathbb{R})$ iff a) and b) hold.

Proof. Sufficiency is easily seen in Fourier transform, by noticing that $\cos|k|t - 1$, $(1+|k|)\sin|k|t/|k|$ and $(1+|k|)^2|k|^{-2}(\cos|k|t - 1)$ are multipliers of L^2 continuous in t.

For the necessity, we note that $L_2(t) \in C^0(L^2, \mathbb{R})$ implies that also $\int_0^t d\tau L_2(\tau) \in C^0(L^2, \mathbb{R})$ and therefore

$$L_1(t) + \int_0^t d\tau L_2(\tau) = -t\tilde{\psi} \in L^2, \quad \text{i.e. } \tilde{\psi} \in L^2.$$

Hence, $|k|^{-1}\sin|k|t\ \tilde{\psi} \in C^0(H^1, \mathbb{R})$ and the condition on $L_1(t)$ yields

$$f(k, t) = |k|^{-2}(1 - \cos|k|t)\tilde{h}(k) \in C^0(H^1, \mathbb{R}), \tag{5.19}$$

which in turn implies

$$(|k|^{-2}\sin|k| - |k|^{-1})\tilde{h} = \int_0^t d\tau f(k, \tau) \in C^0(H^1, \mathbb{R}). \tag{5.20}$$

Finally, the two estimates

$$\tfrac{1}{4}t^2|\tilde{h}(k)| \leq |k|^{-2}(\cos|k|t - 1)|\tilde{h}|,$$

for $|k| \leq 2, t$ sufficiently small, and

$$\tfrac{1}{2}|k|^{-1}|\tilde{h}(k)| \leq (|k|^{-2}\sin|k| - |k|^{-1})|\tilde{h}|,$$

for $|k| \geq 2$, imply $|\tilde{h}|(1 + |k|^2)^{-1/2} \in L^2$, by (5.19), (5.20).

Remarks. The conclusions of the above theorem hold in the more general case in which the condition $\varphi_0 \in L^\infty(\mathbb{R}^s)$ is replaced by that of φ_0 being such that i) and ii), i.e. (5.10) and (5.11), hold; in this case φ_0 is said to be a *regular point* (or admissible) with respect to U, (see Ref. II).

The conditions (5.2), (5.3) characterize those initial data for which the time derivative preserves some sort of localization; in particular (5.3) says that the time derivative of the second component is H^{-1} localized.

The HSSs defined by the absolute minima of the potential are the analogs of the *vacuum sectors* of quantum field theory; those defined by relative minima of U are the analogs of the *false vacuum* sectors[22] and are classically stable (no tunneling).

The solutions of (4.6) which correspond to initial data u_0 satisfying (5.2), (5.3) will be briefly called *generalized stationary solutions*. In general, a sector $\mathcal{H}_{u_0}$ identified by a generalized stationary solution does not contain static solutions; a necessary and sufficient condition is that the elliptic equation

$$\Delta\chi - G'_{\varphi_0}(x, \chi) = -h(\varphi_0)$$

with $h(\varphi_0) \equiv \Delta\varphi_0 - U'(\varphi_0) \in H^{-1}(\mathbb{R}^s)$, has solutions $\chi \in H^1(\mathbb{R}^s)$.

The occurrence of *disjoint* Hilbert structures, stable under time evolution, associated with generalized stationary solutions is a rather remarkable feature in a fully non-linear problem without any approximation or linearization being involved. In a certain sense the generalized stationary solutions play a hierarchical role and exhibit some sort of stability property since they keep their $H^1 \oplus L^2$ perturbations steadily trapped around them. A non-linear structure characterizes the labeling of the sectors by the generalized stationary solutions, since the corresponding initial data do not have a linear structure; however, within a given sector $\mathcal{H}_{\varphi_0}$ all the initial data are described by the affine space generated by φ_0 through $H^1 \oplus L^2$. In general, the time evolution is not described by a linear operator on $\mathcal{H}_{\varphi_0}$.

It is worthwhile to remark that the emergence of disjoint stable structures in the set of solutions of the non-linear equation (4.6) has been made possible by the framework adopted in Chap. 4, in which the Cauchy data were not restricted to be in $H^1 \oplus L^2$. In that case one would have only gotten the sector corresponding to the *trivial vacuum* $\varphi_0 = 0$, $\psi_0 = 0$.[23]

The occurrence of Hilbert space sectors in the set solutions of non-linear field equations allows to establish strong connections with quantum mechanical structures and to recover the analog of quantum mechanical phenomena like linear representations of groups, spontaneous symmetry breaking, pure phases, superselection rules, etc., at the level of classical equations.[24]

The occurrence of disjoint physical worlds or phases is a typical feature of infinitely extended systems, like e.g. those defined by the thermodynamical limit in Statistical Mechanics, where one cannot go from one phase to another by essentially local operations.[25]

---

[22] S. Coleman, Phys. Rev. D **15**, 2929 (1977).

[23] See footnotes 11, 12.

[24] F. Strocchi, Lectures at the Workshop on *Recent Advances in the Theory of Evolution Equations*, ICTP Trieste 1979, published in *Topics in Functional Analysis 1980-81*, Scuola Normale Superiore, Pisa 1982; contribution to the Workshop on *Hyperbolic Equations* (1987), published in Rend. Sem. Mat. Univ. Pol. Torino, Fascicolo speciale 1988, pp. 231-250.

[25] The physical relevance of locality has been emphasized by R. Haag and D. Kastler, J. Math. Phys. **5**, 848 (1964), see also R. Haag, loc. cit. (see footnote 17).

6 Stability under Space Translations. Positive Energy

In this Chapter we discuss the requirements III and IV stated in Chapter 5.

In the sequel we shall denote by $\mathcal{H}_{\varphi_0}$ the Hilbert space sector (HSS) defined by a $\varphi_0 \in L^\infty(\mathbb{R}^s)$ satisfying (5.3), taking always for granted that $\psi_0 \in L^2(\mathbb{R}^s)$.

Stability under space translations means that if $u(x,t) \in \mathcal{H}_{\varphi_0}$ so does $u_a(x,t) \equiv u(x+a,t)$, $\forall a \in \mathbb{R}^s$. Clearly, by the above characterization of Hilbert sectors, such a condition is equivalent to the condition that $\mathcal{H}_{\varphi_{0\,a}} = \mathcal{H}_{\varphi_0}$, $\forall a \in \mathbb{R}^s$, $\varphi_{0\,a}(x) \equiv \varphi_0(x+a)$.

Proposition 6.1. *The Hilbert sector $\mathcal{H}_{\varphi_0}$ is stable under space translations if*

$$\nabla\varphi_0 \in L^2(\mathbb{R}^s). \tag{6.1}$$

Furthermore, the momentum density

$$\mathcal{P}(\varphi,\psi) = \psi\nabla\varphi, \quad \psi \in L^2(\mathbb{R}^s), \quad \varphi - \varphi_0 \in H^1(\mathbb{R}^s)$$

is integrable for any element of $\mathcal{H}_{\varphi_0}$, equivalently its integral $\mathbf{P}$ defines a linear operator $\tilde{\mathbf{P}}$ in $\mathcal{H}_{\varphi_0}$, which acts as the generator of translations

$$\tilde{P}_i u = \{u, P_i\} = \nabla_i u, \tag{6.2}$$

($\{\,,\,\}$ denotes the Poisson bracket), if and only if (6.1) holds.

Proof. The proof of the first statement, namely that $\delta\varphi_0 \equiv \varphi_0(x+a)-\varphi(x) \in H^1(\mathbb{R}^s)$, reduces to the proof that $\delta\varphi_0 \in L^2(\mathbb{R}^s)$ if (6.1) holds. In fact, as a distribution on $C_0^\infty(\mathbb{R}^2)$,

$$\delta\varphi_0 = \int_0^1 d\sigma\, a_i \nabla_i \varphi_0(x+\sigma a)$$

defines a functional which satisfies

$$|\delta\varphi_0(\chi)| \le |a|\, \|\nabla\varphi_0\|_{L^2}\, \|\chi\|_{L^2}$$

as a consequence of (6.1). Thus, it has a continuous extension to $L^2(\mathbb{R}^s)$ and by the Riesz representation theorem it defines an element of $L^2(\mathbb{R}^s)$.

F. Strocchi: *Stability under Space Translations. Positive Energy*, Lect. Notes Phys. **732**, 29–32 (2008)
DOI 10.1007/978-3-540-73593-9_6   © Springer-Verlag Berlin Heidelberg 2008

For the proof of the second statement, since ψ may be an arbitrary element of $L^2(\mathbb{R}^s)$, $\mathcal{P}$ is integrable provided $\nabla\varphi \in L^2(\mathbb{R}^s)$, i.e. $\nabla\varphi_0 \in L^2(\mathbb{R}^s)$, since $\chi = \varphi - \varphi_0 \in H^1(\mathbb{R}^s)$.

The curly brackets in (6.2) defines a linear operator $\tilde{P}$ which is well defined in $\mathcal{H}_\varphi$ and generates the space translations iff (6.1) holds.

Clearly, the possibility of using solutions of non-linear field equations for the description of physical systems requires that such solutions have finite energy-momentum, and the localization properties of the physical measurements require the existence of an energy-momentum density.

The conventional expression of the energy density for the theory described by (4.6) is

$$\mathcal{E}(\varphi, \psi) = \tfrac{1}{2}\left[(\nabla\varphi)^2 + \psi^2\right] + U(\varphi). \qquad (6.3)$$

However, if one adds any function of x, the (Hamilton) equations of motion will remain unchanged and the new expression of the total energy is still formally conserved.

This ambiguity is related to the fact that only energy differences have a physical meaning, so that the concept of finite energy solutions must necessarily make reference to some chosen reference solution. Such a fixing of the energy scale will generally depend on the sector, since $\mathcal{E}(\varphi, \psi)$ is locally but in general not globally integrable. The fixing of the energy scale corresponds to the so-called *infinite volume renormalization* which occurs in the treatment of infinitely extended systems.

Thus, given an Hilbert space sector $\mathcal{H}_{\varphi_0}$, one is led to define a *renormalized energy density* (without loss of generality we can take $\psi_0 = 0$)

$$\mathcal{E}_{ren}(\varphi, \psi) \equiv \mathcal{E}(\varphi, \psi) - \mathcal{E}(\varphi_0, 0)$$

$$= \tfrac{1}{2}\left[(\nabla\chi)^2 + \psi^2\right] + \nabla\chi\nabla\varphi_0 + G(\chi) + U'(\varphi_0)\chi, \qquad (6.4)$$

where $\chi = \varphi - \varphi_0$ and $G(\chi)$ is defined by (5.9).

The background subtraction is, however, not enough for assuring that the renormalized density is globally integrable. The most which can be said, without additional assumptions, is that $\mathcal{E}_{ren}$ is integrable if χ is of compact support and that it identifies an energy functional defined on the whole HSS by a suitable extension[26]. In general, however, the so extended functional will not be the integral over a density and therefore the concept of local energy is problematic. Such a difficulty does not arise if the HSS is defined by a $\varphi_0 \in L^\infty(\mathbb{R}^s)$ with $\nabla\varphi_0 \in L^2(\mathbb{R}^s)$.

Proposition 6.2. [27] *Given a Hilbert space sector defined by a $\varphi_0 \in L^\infty(\mathbb{R}^s)$, a (renormalized) energy density can be defined on it with a convergent infinite volume integral if*

$$\nabla\varphi_0 \in L^2(\mathbb{R}^s). \qquad (6.5)$$

---

[26] Ref. II quoted in footnote 5.
[27] See Ref. II.

Proof. By Lemma 5.3 $G'(\chi)$ is globally Lipschitz continuous and therefore $G'(\chi) \in L^2(\mathbb{R}^s), \forall \chi \in H^1(\mathbb{R}^s)$. Now, from the identity

$$G(\chi_1) - G(\chi_2) = \int_0^1 d\sigma \frac{d}{d\sigma} G(\chi_1 + \sigma(\chi_2 - \chi_1)) =$$

$$= \int_0^1 d\sigma(\chi_2 - \chi_1) G'(\chi_1 + \sigma(\chi_2 - \chi_1)),$$

one has

$$\int d^s x |G(\chi_1) - G(\chi_2)| \leq \sup_{0 \leq \sigma \leq 1} \|G'(\chi_1 + \sigma(\chi_2 - \chi_1))\|_{L^2} \|\chi_2 - \chi_1\|_{L^2}$$

and, since $G(0) = 0$, $G(\chi) \in L^1(\mathbb{R}^s)$.

On the other hand,

$$\nabla \chi \nabla \varphi_0 + U'(\varphi_0)\chi = \nabla(\chi \nabla \varphi_0) - h(\varphi_0)\chi$$

and the second term on the r.h.s. is integrable since $h \in H^{-1}(\mathbb{R}^s), \chi \in H^1(\mathbb{R}^s)$. By (6.5), $\chi \nabla \varphi_0 \in L^1(\mathbb{R}^s)$ and therefore the infinite volume limit of the integral of the first term vanishes. The other terms in (6.4) are clearly integrable.

No renormalization is needed for the momentum, since without loss of generality we can take $\psi_0 = 0$ and the background momentum subtraction vanishes.

It is not difficult to show[28] that the infinite volume integrals of the renormalized energy-momentum densities define conserved quantities and that the corresponding functionals are continuous in the Hilbert space topology of the HSS, if (6.5) holds.

A HSS defined by a $\varphi_0 \in L^\infty(\mathbb{R}^s)$ with $\nabla \varphi_0 \in L^2(\mathbb{R}^s)$ will be called a *Hilbert space sector with energy-momentum density* or briefly a *physical sector*.

A related question is the stability of a sector under external perturbations and an important role is played by condition IV of Chapter 5, namely that the (renormalized) energy be bounded from below. Now, even if the potential is bounded from below, in general the renormalized energy may not be so.

Proposition 6.3. *The renormalized energy is bounded from below in the HSSs defined by absolute minima of the potential (***vacuum sectors***)*

Proof. In fact, in (6.4) $\nabla \varphi_0 = 0$ and, since φ_0 is an absolute minimum

$$G(\chi) + U'(\varphi_0)\chi = U(\varphi_0 + \chi) - U(\varphi_0) \geq 0.$$

---

[28] See Ref. II.

The energy is not bounded from below in the sectors defined by relative minima of the potential (false vacuum sectors) and one expects instability against external field perturbations.

In conclusion, the set of solutions of the non-linear field equation (4.6) which have a reasonable physical interpretation are those belonging to Hilbert space sectors with energy-momentum density (*physical sectors*), and a distinguished role is played by the vacuum sectors . (For time-independent solutions defining physical sectors, see Appendix E). The analogy with the corresponding structures in quantum field theory[29] is rather remarkable.

---

[29] See references in footnotes 16 and 17.

7 Noether Theorem and Symmetry Breaking

The existence of sectors, i.e. of "disjoint physical worlds" in the set of solutions of the non-linear equation (4.6), provides the mathematical and physical basis for the mechanism of spontaneous symmetry breaking briefly discussed in Chap. 2. We can now understand the relation between the Noether theorem, the existence of conserved currents and the occurrence of spontaneous symmetry breaking, which, among other things, implies the lack of existence of the corresponding generators. As shown by the following Propositions, the mechanism of spontaneous symmetry breaking is related to the instability of a physical world under a symmetry operation.

Proposition 7.1. [30] *Let G denote the group of internal symmetries of (4.6). Then*

1) G maps sectors into sectors and HSS into HSS

$$G : \mathcal{H}_\varphi \to \mathcal{H}_{g(\varphi)}, \quad \forall g \in G,$$

giving rise to orbits of sectors and of HSS.
2) Each HSS $\mathcal{H}_\varphi$ determines a subgroup G_φ of G such that

$$G_\varphi : \mathcal{H}_\varphi \to \mathcal{H}_\varphi.$$

*G_φ is called the **stability group** of $\mathcal{H}_\varphi$ and $\mathcal{H}_\varphi$ is the carrier of a representation of its stability group.*
3) A necessary and sufficient condition for G_φ being the stability group of $\mathcal{H}_\varphi$ is that there exists one element $\bar{\varphi} \in \mathcal{H}_\varphi$ such that

$$G_\varphi \bar{\varphi} \in \mathcal{H}_\varphi. \tag{7.1}$$

---

[30] Ref. II.

F. Strocchi: *Noether Theorem and Symmetry Breaking*, Lect. Notes Phys. **732**, 33–37 (2008)
DOI 10.1007/978-3-540-73593-9_7   © Springer-Verlag Berlin Heidelberg 2008

Proof. By the characterization of internal symmetries given by (3.6), (3.7), $u'(t) - u(t) \in C^0(H^1 \oplus L^2, \mathbb{R})$ implies

$$g(\varphi'(t)) - g(\varphi(t)) = A_g(\varphi'(t) - \varphi(t)) \in C^0(H^1 \oplus L^2, \mathbb{R}),$$

so that sectors are mapped into sectors.

Furthermore, if $u_0 = \{\varphi_0, \psi_0\}$, with $\varphi_0 \in L^\infty(\mathbb{R}^s)$, $\psi_0 \in L^2(\mathbb{R}^s)$, satisfies condition b) of Theorem 5.2, it follows that $A_g\varphi_0 + a_g \in L^\infty(\mathbb{R}^s)$, $A_g\psi_0 \in L^2(\mathbb{R}^s)$ and, by (3.6), (3.8),

$$\Delta g(\varphi_0) - U'(g(\varphi_0)) = A_g(\Delta\varphi_0 - U'(\varphi_0)) \in H^{-1}(\mathbb{R}^s),$$

i.e. g maps HSS into HSS.

Finally, for any element φ of $\mathcal{H}_{\varphi_0}$, putting $\chi = \varphi - \varphi_0$, one has

$$g(\varphi) - \varphi_0 = A_g\chi + g(\varphi_0) - \varphi_0 \tag{7.2}$$

and since for any $g \in G_{\varphi_0}$, $g(\varphi_0) - \varphi_0 \in H^1(\mathbb{R}^s) \oplus L^2(\mathbb{R}^s)$, by (7.2) the mapping g induces an affine mapping on $H^1 \oplus L^2$ to which $\mathcal{H}_{\varphi_0}$ is naturally identified, by Theorem 5.2.

Conversely, by arguing as for (7.2), if $\exists \bar{\varphi} \in \mathcal{H}_{\varphi_0}$ such that $g(\bar{\varphi}) - \bar{\varphi} \in H^1 \oplus L^2$ so does $g(\varphi) - \varphi_0$, i.e. $g \in G_\varphi$.

Since, as discussed before, different HSSs define "disjoint physical worlds", an internal symmetry of the field equation (4.6) gives rise to a symmetry of the physical world described by the Hilbert sector $\mathcal{H}_\varphi$ only if it maps $\mathcal{H}_\varphi$ into $\mathcal{H}_\varphi$. Otherwise the symmetry is *spontaneously broken*.

As discussed in the Introduction, if $\mathcal{H}_\varphi$ is not stable under G, its elements cannot be classified in terms of irreducible representations of G. It is now clear what distinguishes the infinite dimensional case with respect to the finite dimensional one. In the latter case, degenerate ground states related by a continuous symmetry, cannot be separated by potential barriers and one can move from one to the other by physically realizable operations. In the infinite dimensional case, degenerate ground states characterize different large distance behaviours of the field configurations, so that, even if they are related by a continuous symmetry, they cannot be related by physically realizable operations, since the latter ones must both involve finite energy *and* be essentially localized.

When the field equations can be derived by a Lagrangean, the link between the invariance group of the Lagrangean and the existence of conservation laws is provided by the classical Noether's theorem. The existence of a continuity equation or a local conservation law, however, does not in general imply the existence of a *conserved charge*, since, first of all, the integral which defines the charge

$$Q^i = \int d^3x \, J_0^i(x)$$

may not converge.

Thus, the standard accounts of the Noether theorem implicitly apply to the solutions which decrease sufficiently fast at infinity, i.e. essentially to the "trivial vacuum" sector $\mathcal{H}_{\varphi=0}$.

In the general case when the solutions do not belong to $H^1 \oplus L^2$, a criterium for the existence of a conserved charge and a corresponding linear operator in $\mathcal{H}_\varphi$, acting as the generator of the corresponding symmetry, is provided by the following version of Noether theorem.[31] Again, the structure of Hilbert space sectors provides a simple solution of the problem of compatibility of Noether theorem and symmetry breaking.

For simplicity, we consider the case of real fields and of linear transformations with $a_g = 0, \lambda_g = 1$, the generalization being straightforward.

Theorem 7.2. *Let G be an N-parameter continuous (Lie) group of internal symmetries of the field equation (4.6) (or of the Lagrangean from which they are derived), then there exist N currents $J^i_\mu(u(x,t)) \equiv J^i_\mu(x,t)$, which obey the continuity equation*

$$\partial^\mu J^i_\mu(x,t) = 0, \qquad i = 1, ...N \tag{7.3}$$

(local conservation law).

Given a physical HSS $\mathcal{H}_{\varphi_0}$, and a one-parameter subgroup $G^{(j)} \subset G$, the Noether charge

$$Q^j(u(t)) \equiv \int d^s x \, J^j_0(u(x,t)) \tag{7.4}$$

exists and is independent of time for all solutions $u(x,t) \in \mathcal{H}_{\varphi_0}$, equivalently it defines a linear operator $\tilde{Q}^j : \mathcal{H}_{\varphi_0} \to \mathcal{H}_{\varphi_0}$, acting as the generator of the corresponding symmetry transformation

$$\tilde{Q}^j u \equiv \{u, Q^j\} = \delta^{(j)} u, \tag{7.5}$$

where the curly brackets denote the Poisson brackets, iff $G^{(j)}$ is a subgroup of the stability group G_{φ_0} of $\mathcal{H}_{\varphi_0}$.

Furthermore, in this case the subgroup $G^{(j)}$ is represented by unitary operators in $\mathcal{H}_{\varphi_0}$.

Proof. We omit the proof of the first part, which is standard and can be found in any textbook of classical field theory (see, e.g., the references given for Theorem 3.2).

For the second part, we start by discussing the convergence of the integral (7.4). The stability of $\mathcal{H}_{\varphi_0}$ under $G^{(j)}$ is equivalent to its stability under infinitesimal transformations of $G^{(j)}$

$$\varphi \to \varphi + \epsilon^{(j)} \, \delta^{(j)} \varphi, \qquad \delta^{(j)}\varphi = \frac{\partial}{\partial \epsilon^{(j)}} A_{g_\epsilon} \varphi|_{\epsilon^{(j)}=0},$$

namely to the condition $\delta^{(j)}\varphi \in H^1(\mathbb{R}^s)$.

---

[31] F. Strocchi, loc.cit. (see footnote 24).

Now, $J_0^j(\varphi, \psi) = \psi \, \delta^{(j)} \varphi$ and therefore, since ψ may be an arbitrary element of $L^2(\mathbb{R}^s)$, $J_0^j \in L^1(\mathbb{R}^s)$ iff $\delta^{(j)} \varphi \in L^2(\mathbb{R}^s)$. On the other hand, for a physical Hilbert space sector (see Chap. 6), $\nabla \varphi \in L^2(\mathbb{R}^s)$, which implies $\nabla A_g(\varphi) = A_g \nabla \varphi \in L^2(\mathbb{R}^s)$ and therefore $\nabla \delta^{(j)} \varphi = \delta^{(j)} \nabla \varphi \in L^2(\mathbb{R}^s)$. Hence, for a physical sector $\delta^{(j)} \varphi \in L^2(\mathbb{R}^s)$ is equivalent to $\delta^{(j)} \varphi \in H^1(\mathbb{R}^s)$.

Equivalently (it is easy to see that the Poisson brackets of a bilinear form $Q = \int d^s x \, \psi A \varphi$ defines an operator $\tilde{Q} : \mathcal{H}_\varphi \to \mathcal{H}_\varphi$ iff the integral converges for any $u = (\varphi, \psi) \in \mathcal{H}_\varphi$), (7.5) defines a linear operator $\tilde{Q}^j : \mathcal{H}_\varphi \to \mathcal{H}_\varphi$, which acts as the generator of the symmetry transformation iff $\delta^{(j)} \varphi \in L^2(\mathbb{R}^s)$.

In this case, the formal series

$$\exp\left(\alpha \tilde{Q}^j\right) u = u + \alpha \{u, Q^j\} + \tfrac{1}{2} \alpha^2 \left\{\{u, Q^j\}, Q^j\right\} + ..., \quad \alpha \in \mathbb{R},$$

has the following properties: i) all terms belong to $\mathcal{H}_{\varphi_0}$ and ii) it satisfies

$$(d/d\alpha) \exp\left(\alpha \tilde{Q}^j\right) u = \{\exp\left(\alpha \tilde{Q}^j\right) u, Q^j\}, \quad \exp\left(\alpha \tilde{Q}^j\right) u|_{\alpha=0} = u,$$

$$\left(\exp\left(\alpha \tilde{Q}^j\right) u_1, \exp\left(\alpha \tilde{Q}^j\right) u_2\right) = (u_1, u_2),$$

with $(\,,\,)$ the scalar product in $\mathcal{H}_{\varphi_0}$. Thus, the series defines a unitary operator in $\mathcal{H}_{\varphi_0}$ which implements the given symmetry.

For the time independence of the charge integral, we recall that it is related to the continuity equation of the current J_μ^i by the following argument. One integrates $\partial^\mu J_\mu^i(x, t) = 0$ over the space-time region $\mathcal{V} \equiv \{x \in V = \text{a bounded space volume}, \, t \in [0, \tau]\}$ and uses Gauss theorem to get

$$0 = \int_\mathcal{V} d^s x \, dt \, \partial^\mu J_\mu^i(x, t) = Q_V^i(\tau) - Q_V^i(0) + \Phi_S(\boldsymbol{J}^{(i)}), \qquad (7.6)$$

where $\Phi_S(\boldsymbol{J}^i)$ is the flux of $\boldsymbol{J}^{(i)} = \nabla \varphi \, \delta^{(i)} \varphi$ over the boundary surface $S \equiv \{x \in \partial V, \, t \in [0, \tau]\}$. The time independence of the charge integral is then equivalent to the vanishing of the flux $\Phi_S(\boldsymbol{J})$ in the limit $V \to \infty$. Since $\boldsymbol{J}^{(j)} = \nabla \varphi \, \delta^j \varphi = \nabla \chi \, (\delta^j \varphi_0 + \delta^j \chi)$ the flux vanishes $\forall \nabla \chi \in L^2$, iff $\delta^j \varphi_0 = 0$. Thus, $\tilde{Q}_V^j(\tau) - \tilde{Q}_V^j(0)$ converges to zero in $\mathcal{H}_\varphi$ iff $\delta^j \varphi_0 = 0$.

Remark 1. It is not difficult to find the analog of the above theorem in the more general case of non-internal symmetries, which commute with time evolution.

Remark 2. The notion of physical Hilbert space sector clarifies the conditions for the existence of a time independent linear operator, which generates the symmetry, and accounts for the mechanism of spontaneous symmetry breaking compatibly with Noether theorem.

As shown by the above discussion, in general the continuity equation for J^i_μ may fail to give rise to a conserved charge by a mechanism which is closely related to that of symmetry breaking in quantum systems.[32][33]

For the vanishing of the flux at infinity, a crucial role is played by the condition $\nabla \varphi_0 \in L^2(\mathbb{R}^s)$, which is related to the invariance of the Hilbert sector under space translations and characterizes the physical sectors. This is no longer the case in gauge field theories, since the energy-momentum density involves the covariant derivative $(\nabla + A)\varphi$ (where A denotes the gauge potential), rather than $\nabla \varphi$. This opens the way to the Higgs mechanism of symmetry breaking [34] for which the boundary effects give rise to a charge leaking at infinity[35].

[32] J. Goldstone, Nuovo Cim. **19**, 154 (1961); J. Goldstone, A. Salam and S. Weinberg, Phys. Rev. **127**, 965 (1962); Y. Nambu and G. Jona-Lasinio, Phys. Rev. **122**, 345 (1961); **124**, 246 (1961).

[33] For a simple account see F. Strocchi, *Elements of Quantum Mechanics of Infinite Systems*, World Scientific 1985.

[34] P.W. Higgs, Phys. Lett. **12**, 132 (1964); T.W. Kibble, *Proc. Oxford Int. Conf. Elementary Particles*, Oxford, Oxford Univ. Press 1965; G.S. Guralnik, C.R. Hagen and T.W. Kibble, in *Advances in Particle Physics* Vol. 2, R.L. Cool and R.E. Marshak eds., Interscience, New York, 1968 and refs. therein. See also the references in the footnote below.

[35] G. Morchio and F. Strocchi, in *Fundamental Problems of Gauge Field Theory*, G. Velo and A.S. Wightman eds., Plenum 1986; F. Strocchi, in *Fundamental Aspects of Quantum Theory*, V. Gorini and A. Frigerio eds., Plenum 1986.

8 Examples

1) Non-linear Scalar Field in One Space Dimension

The model describes the simplest non-linear field theory and it can be regarded as a prototype of field theories in one space dimension ($s = 1$). The model can also be interpreted as a non-linear generalization of the wave equation. The interest of the model is that, even at the classical level, it has stable solutions with a possible particle interpretation[36].

The model is defined by the potential

$$U = -\tfrac{1}{2}m^2\varphi^2 + \tfrac{1}{4}\lambda\varphi^4 = \tfrac{1}{4}\lambda(\varphi^2 - \mu^2)^2 - \tfrac{1}{4}\lambda\mu^4, \quad \mu^2 \equiv m^2/\lambda, \qquad (8.1)$$

and therefore the equations of motion read

$$\Box\varphi = -\lambda\varphi(\varphi^2 - \mu^2). \qquad (8.2)$$

i) Vacuum state solutions

The simplest solutions are the *ground state solutions*, invariant under space and time translations, i.e. $\varphi = const$. If the field φ takes values in $\mathbb{R}$, there are only three possibilities

$$\varphi_0^{\pm} = \pm\mu, \qquad \varphi_0 = 0. \qquad (8.3)$$

By the discussion of Chaps. 5–7, $\varphi_0^{\pm}$ define disjoint Hilbert space sectors $\mathcal{H}_{\pm}$, for which an energy-momentum density can be defined and for which the energy is bounded below. The other constant solution $\varphi_0 = 0$, corresponding to the so-called trivial vacuum sector, still defines a Hilbert space sector with energy-momentum density, but the energy is not bounded below and therefore in this case the sector is not energetically stable under external perturbations (see Chap. 7). This would be the only vacuum state solution available in Segal's approach.

If the field φ takes values in $\mathbb{R}^n, n > 1$, the internal symmetry group is the continuous group G of transformations (3.6), (3.7) with $\lambda = 1, a = 0$. In this case, besides the trivial vacuum solution $\varphi_0 = 0$, the non-trivial vacuum

---

[36] J. Goldstone and R. Jackiw, Phys. Rev. **D11**, 1486 (1975). See also R. Rajaraman, *Solitons and Instantons*, North-Holland 1982 and references therein.

F. Strocchi: *Examples*, Lect. Notes Phys. **732**, 39–43 (2008)
DOI 10.1007/978-3-540-73593-9_8 © Springer-Verlag Berlin Heidelberg 2008

solutions are given by the points of the orbit

$$\{\varphi_0^g \equiv A_g \bar{\varphi}_0, \ \ g \in G, \ \ \bar{\varphi}_0^2 = \mu^2\}. \tag{8.4}$$

For $n = 1$, the internal symmetry group is the discrete group

$$Z_2 : \varphi \rightarrow -\varphi.$$

Clearly, in all cases, the internal symmetry group is unbroken in the trivial vacuum sector $\mathcal{H}_0$, but it is spontaneously broken in each "pure phase" $\mathcal{H}_g$, defined by φ_0^g.

ii) *Time independent solutions defining physical Hilbert space sectors. Kinks* Another interesting class are the time independent solutions, which satisfy

$$(\partial_x)^2 \varphi = \lambda \varphi(\varphi^2 - \mu^2). \tag{8.5}$$

This implies

$$\partial_x(\tfrac{1}{2}\varphi_x^2 - \tfrac{1}{4}\lambda(\varphi^2 - \mu^2)^2) = 0, \ \ \ \varphi_x \equiv \partial_x \varphi,$$

i.e.

$$\tfrac{1}{2}\varphi_x^2 = \tfrac{1}{4}\lambda(\varphi^2 - \mu^2)^2 + C, \ \ \ C = \text{constant}. \tag{8.6}$$

For simplicity, we consider the case in which φ takes values in $\mathbb{R}$, leaving the straightforward generalization as an exercise.

The discussion of the solutions of (8.5), as given in the literature, (see e.g. the references in the previous footnote), is done under the condition that they have finite energy when the potential is so renormalized that it vanishes at its absolute minimum. This means that

$$\tfrac{1}{2}(\nabla\varphi)^2 + \tfrac{1}{4}\lambda(\varphi^2 - \mu^2)^2 \in L^1.$$

By the discussion of Chap. 5, this appears as too restrictive, since it does not consider the possibility of energy renormalization, (6.4), and in particular it crucially depends on the overall scale of the potential (it also excludes the trivial vacuum solution $\varphi_0 = 0!$). For these reasons we prefer to leave open the energy renormalization.

To simplify the discussion we will only assume that φ has (bounded) limits $\varphi(\pm\infty)$, when $x \rightarrow \pm\infty$ (regularity at infinity). Then, quite generally, since U is by assumption of class C^2, also $U'(\varphi)$ has bounded limits as $x \rightarrow \pm\infty$ and (8.5) implies that $d^2\varphi/dx^2$ also does. On the other hand, for any test function f of compact support, with $\int f(x)dx = 1$,

$$\lim_{a\rightarrow\pm\infty}(\Delta\varphi)(x + a) = \lim_{a\rightarrow\pm\infty}\int \Delta\varphi(x + a) \, f(x)dx$$

$$= \lim_{a\rightarrow\pm\infty}\int \varphi(x + a)\Delta f(x)dx = \varphi(\pm\infty)\int \Delta f(x) \, dx = 0.$$

Then, (8.5) implies
$$U'(\varphi(\pm\infty)) = 0. \tag{8.7}$$

Now, for physical sectors $\nabla\varphi \in L^2$, so that, unless $\varphi(\pm\infty) = 0$, the constant C in (8.6) must vanish and one has

$$\varphi_x(x) = \varepsilon(x)\sqrt{\lambda/2}\,(\varphi^2(x) - \mu^2), \tag{8.8}$$

with $\varepsilon(x)^2 = 1$. Actually, (8.5) implies that $\varepsilon(x)$ is independent of x, i.e. $\varepsilon(x) = \pm 1$. Equation (8.8) can easily be integrated and it gives

$$\varphi(x) = \mp\mu\tanh(\sqrt{\lambda/2}\,\mu(x - a)), \tag{8.9}$$

where a is an integration constant.

The plus/minus sign gives the so-called *kink/anti-kink* solution, respectively. Such solutions do not vanish at $x \to \pm\infty$, but, nevertheless, they have some kind of localization, since they significantly differ from the constants φ_0^+, φ_0^- only in a region of width $(\sqrt{\lambda}\mu)^{-1}$. They are not local perturbations of the ground state solutions $\varphi_0^\pm$ and in fact they define different Hilbert sectors $\mathcal{H}_k, \mathcal{H}_{\bar{k}}$. The corresponding renormalized energy momentum density is defined by

$$\mathcal{E}_{ren} = \tfrac{1}{2}(\nabla\varphi)^2 + U(\varphi) + \tfrac{1}{4}\lambda\mu^4 = \tfrac{1}{2}(\nabla\varphi)^2 + \tfrac{1}{4}\lambda(\varphi^2 - \mu^2)^2$$

and it is localized around the "centre of mass" of the kink, namely $x = a$. (It is instructive to draw the shape of the kink solution). The total renormalized energy is

$$E_k = \tfrac{2}{3}\sqrt{2}\,m^3/\lambda \tag{8.10}$$

and it clearly exhibits the non-perturbative nature of the kink solution.

iii) *Moving kink. Particle behaviour*
Since (8.2) is invariant under a Lorentz transformation

$$x \to x' = (x - vt)/\sqrt{1 - v^2}, \quad t \to t' = (t - vx)/\sqrt{1 - v^2},$$

(where the velocity of light c is put $= 1$) if $\varphi(x, t)$ is a solution, so is $\varphi'(x, t) \equiv \varphi(x', t')$. Thus, from the static solutions (8.9) we can generate time dependent ones (for simplicity we put $a = 0$)

$$\varphi(x, t) = \mp\mu\tanh(\sqrt{\lambda/2}\,\mu(x - vt)/\sqrt{1 - v^2}), \quad v^2 < 1. \tag{8.11}$$

The energy-momentum density is localized around the point $x = vt$ ("center of mass" of the kink), which moves with velocity v (*moving kink* solution).

Clearly, $\varphi(x, t) - \varphi(x, 0) \in C^\circ(H^1, \mathbb{R})$, i.e. $\varphi(x, t)$ defines a sector. Furthermore $\varphi(x, 0) \in L^\infty(\mathbb{R})$, $\psi(x, 0) = \dot{\varphi}(x, 0) \in L^2(\mathbb{R})$ and obviously condition b) of Theorem 5.1 is satisfied; then $(\varphi(x, 0), \psi(x, 0))$ defines a Hilbert space sector.

This implies the stability of such solutions under $H^1 \oplus L^2$ perturbations (see Chap. 5). This settles the problem of stability of the kink sector and, thanks to Theorem 5.1, the proof does not involve expansions or linearizations[37]. It is not difficult to see that the static kink solution, corresponding to $v = 0$ in (8.11), belongs to the same sector defined by the corresponding moving kink solution.

From a physical point of view (energy-momentum localization and stability), the kink is a candidate to describe particle-like excitations associated with (8.2). In fact, in the past this feature has motivated attempts to use such kink-like solution as a non-perturbative semi-classical approach to the descriptions of baryons in quantum field theory[38].

2) The Sine-Gordon Equation

The Sine-Gordon equation is

$$\Box\varphi = -g\sin\varphi, \tag{8.12}$$

where $\varphi(x,t)$ is a scalar field in one space dimension. It is of great interest in various fields of theoretical physics, like propagation of crystal dislocation, magnetic flux in Josephson lines, Bloch wall motion in magnetic crystals, fermion bosonization in the Thirring model of elementary particle interactions, etc.[39]

i) *Static solutions*
The simplest static solutions are the constants

$$\varphi = \pi n, \quad n \in \mathbb{Z}. \tag{8.13}$$

They all define disjoint Hilbert space sectors and for n even correspond to absolute minima of the potential

$$U = g(1 - \cos\varphi). \tag{8.14}$$

In this case the energy is bounded below in the corresponding Hilbert sectors.

---

[37] See e.g. R. Rajaraman, Phys. Rep. **21**, 227 (1975), especially Chap. 3.2.

[38] R.F. Dashen, B. Hasslacher and A. Neveu, Phys. Rev. **D10**, 4130 (1974); J. Goldstone and R. Jackiw, Phys. Rev. **D11**, 1486 (1975); for a rich collection of important papers see C. Rebbi and G. Soliani, *Solitons and Particles*, World Scientific 1984.

[39] See A. Barone, F. Esposito and C.J. Magee, Theory and Applications of the Sine-Gordon Equation, in Riv. Nuovo Cim. **1**, 227 (1971); A.C. Scott, F.Y. Chiu, and D.W. Mclaughlin, Proc.I.E.E.E. **61**, 1443 (1973); G.B. Whitham, *Linear and Non-Linear Waves*, J. Wiley 1974; S. Coleman, Phys. Rev. **D11**, 2088 (1975); S. Coleman, *Aspects of Symmetry*, Cambridge Univ. Press 1985; J. Fröhlich, in *Invariant Wave Equations*, G. Velo and A.S. Wightman eds., Springer-Verlag 1977.

The internal symmetries of (8.12) are

$$\varphi \to \varphi + 2\pi n$$

and

$$\varphi \to -\varphi.$$

They are broken in the sectors $\mathcal{H}_{\pi n}$ defined by the vacuum solutions (8.13).

To determine other non-trivial static solutions we proceed as in Example 1) The equation

$$\Delta\varphi = g \sin\varphi, \tag{8.15}$$

implies

$$\frac{d}{dx}[\tfrac{1}{2}\varphi_x^2 + g\cos\varphi] = 0, \tag{8.16}$$

i.e.

$$\tfrac{1}{2}\varphi_x^2 + g\cos\varphi = C, \qquad C = \text{constant}. \tag{8.17}$$

As in the previous example, we prefer to leave open the energy renormalization and we classify all the solutions of (8.17) which have (bounded) limits $\varphi_{\pm\infty}$ when $x \to \pm\infty$. By the same argument as before, one finds that $\sin\varphi_{\pm\infty} = 0$, i.e.

$$\varphi_{\pm\infty} = \pi n_{\pm}, \qquad n_{\pm} \in \mathbb{Z} \tag{8.18}$$

and, from the condition $\nabla\varphi \in L^2$, one gets

$$n_+ = n_- \bmod 2\pi, \qquad C = \varepsilon g,$$

with $\varepsilon = 1$ for $n_+ = even$, $\varepsilon = -1$ for $n_+ = odd$. Actually, the case $n_+ = odd$ is ruled out by (8.17), which requires $C - g \cos\varphi = \tfrac{1}{2}\varphi_x^2 \geq 0$. Then

$$\varphi_x(x) = \varepsilon(x)\sqrt{2g}\sqrt{1 - \cos\varphi(x)}, \qquad \varepsilon(x)^2 = 1 \tag{8.19}$$

and again $\varepsilon(x) = \pm 1$, by (8.16).

Equation (8.19) can be easily integrated and it gives

$$\varphi(x) = \pm 4\tan^{-1}[\exp\sqrt{g}(x - a)] \equiv \varphi_{s/\bar{s}} \tag{8.20}$$

with a an integration constant. Corresponding to the $+$ or $-$ sign, the solution is called *soliton* or *anti-soliton*.

ii) *Moving soliton solutions*

As before, moving soliton (or anti-soliton) solutions can be obtained by Lorentz transformations, i.e. by replacing $x - a$ in (8.20) by $(x - a - vt)/\sqrt{1 - v^2}$. A remarkable property of solitons with respect to kinks is that they are unaltered by scattering. The literature on solitons is vast (see e.g. the references in the previous footnote).

It is not difficult to see that φ_s and $\varphi_{\bar{s}}$ define different Hilbert sectors $\mathcal{H}_s, \mathcal{H}_{\bar{s}}$ (also different from the $\mathcal{H}_{\pi n}$, defined by the vacuum solution (8.13)).

9 The Goldstone Theorem

The mechanism of SSB does not only provide a general strategy for unifying the description of apparently different systems, but it also provides information on the energy spectrum of an infinite dimensional system, by means of the so-called *Goldstone theorem*,[40] according to which to each broken generator T of a continuous symmetry there corresponds a massless mode, i.e. a free wave. The quantum version of such a statement has been turned into a theorem,[41] whereas, as far as we know, no analogous theorem has been proved for classical (infinite dimensional) systems and the standard accounts seem to rely on heuristic arguments.

The standard heuristic argument, which actually goes back to Goldstone, considers as a prototype the nonlinear equation (3.1)

$$\Box \varphi + U'(\varphi) = 0,$$

where the multi-component real field φ transforms as a linear representation of a Lie group G and the potential U is invariant under the transformations of G. This implies that for the generator T^α one has

$$0 = \delta^\alpha U(\varphi) = U'_j(\varphi) \, T^\alpha_{jk} \, \varphi_k, \quad \forall \varphi \tag{9.1}$$

and therefore the derivative of this equation at $\varphi = \overline{\varphi}$ gives

$$U''_{jk}(\overline{\varphi}) \, (T^\alpha \overline{\varphi})_k = 0. \tag{9.2}$$

Thus, in an expansion of the potential around $\overline{\varphi}$, the quadratic term, which has the meaning of a mass term, has a zero eigenvalue in the direction $T^\alpha \overline{\varphi}$. This is taken as evidence that there is a massless mode.

In our opinion, the argument is not conclusive since it involves an expansion and one should in some way control the effect of higher order terms; moreover, it is not clear that there are (physically meaningful) solutions in the direction of $T^\alpha \overline{\varphi}$ for all times, so that for them the quadratic term disappears. In any case, the argument does not show that there are massless solutions as in the quantum case.

---

[40] J. Goldstone, Nuovo Cimento **19**, 154 (1961).

[41] J. Goldstone, A. Salam and S. Weinberg, Phys. Rev. **127**, 965 (1962); J. Swieca, *Goldstone's theorem and related topics*, Cargèse lectures 1969.

F. Strocchi: *The Goldstone Theorem*, Lect. Notes Phys. **732**, 45–49 (2008)
DOI 10.1007/978-3-540-73593-9_9   © Springer-Verlag Berlin Heidelberg 2008

Another heuristic argument appeals to the finite dimensional analogy, where the motion of a particle along the bottom of the potential, i.e. along the orbit $\{g^\alpha(\lambda)\overline{\varphi}\}$, where $g^\alpha(\lambda)$, $\lambda \in \mathbb{R}$, is the one-parameter subgroup generated by T^α, does not feel the potential, since $U'(g^\alpha\overline{\varphi}) = 0$, and therefore the motion is like a free motion. This is considered as evidence that, correspondingly, in the infinite dimensional case there are massless modes. Again the argument does not appear complete, since it is not at all clear that there are physically meaningful solutions, i.e. belonging to the physical sector of $\overline{\varphi}$ and therefore of the form $\varphi = \overline{\varphi} + \chi$, $\chi \in H^1(\mathbb{R}^s)$, $s =$ space dimension, of zero mass.

We propose a version[42] of the Goldstone theorem for classical fields as a mathematically acceptable substitute and correction of the above heuristic arguments.

We consider the case of space dimension $s = 3$, unless otherwise stated and for simplicity the case of compact semi-simple Lie group G of internal symmetries. The potential is assumed to be of class C^3.

The argument relies on some basic fact on the asymptotic solutions of (4.6) which we briefly recall for the convenience of the reader.

Given a solution $u(t)$ of the integral equation (4.6), its asymptotic time $(t \to \pm\infty)$ behaviour defines the so-called *scattering configurations* or *asymptotic states* $u_\pm(t)$ associated with $u(t)$.

The behaviour of $f(u)$ near $u = 0$ plays a crucial role for such asymptotic limits and if $f(u) - f'(0)\,u$ vanishes to a sufficiently high degree, e.g. as $O(u^3)$, i) such limits $u_\pm(t)$ exist and ii) their time evolution is that corresponding to the differential operator $\square + f'(0)$, i.e.

$$u_\pm(t') = \mathcal{W}(t' - t)\,u_\pm(t),$$

where $\mathcal{W}(t)$ denotes the propagator corresponding to the differential operator $\square + f'(0)$, (if $f'(0) = 0, \mathcal{W}(t)$ is the free wave equation propagator $W(t)$ defined in Chap. 4).

The mathematical theory of scattering for the nonlinear wave equation is well developed and it is beautifully reviewed by W. Strauss, *Non-linear Wave Equations*, Am. Math. Soc. 1989.

The mathematical problem of the existence of the scattering configurations (the so-called scattering theory) is to guarantee the well definiteness of the Yang-Feldman equations

$$u_\pm(t) = u(t) + \int_t^{\pm\infty} ds\, U_0(t - s)\, f(u(s)), \qquad (9.3)$$

which express $u_\pm(t)$ in terms of the solution $u(t)$ and of the propagator $\mathcal{W}$. The Yang-Feldman equations can be interpreted as a form of the integral (4.6) with initial data given at $t = \pm\infty$, respectively.

---

[42] F. Strocchi, Phys. Lett. **A267**, 40 (2000).

The problem of the existence of the asymptotic limits reduces to estimating the asymptotic time decay of the nonlinear term $f(u(s))$ such that the integrals on the r.h.s. of the Yang-Feldman equations exist. This can be done by using the Basic L^∞ estimates on the time decay of the free solutions (see Strauss' book quoted above, pp. 5-6).

For small amplitude solutions, i.e. for initial data small in some norm, e.g. of the form εu for fixed u, the asymptotic limits are completely governed by the behaviour of $f(u)$ near $u = 0$.

We can now state a classical counterpart of the Goldstone's theorem.

Theorem 9.1. *Let G be an N-parameter continuous (Lie) group of internal symmetries of the nonlinear equation (3.1) and $\mathcal{H}_{\overline{\varphi}}$ the Hilbert Space Sector, defined by an absolute minimum $\overline{\varphi}$ of the potential U, where G is spontaneously broken down to $G_{\overline{\varphi}}$, the stability group of $\overline{\varphi}$.*

Then, for any generator T^α, such that $T^\alpha \overline{\varphi} \neq 0$,

i) *there are scattering configurations, associated to solutions belonging to the sector $\mathcal{H}_{\overline{\varphi}}$, which are solutions of the free wave equation (**Goldstone modes**).*

ii) *for any sphere Ω_R of radius R and any time T there are solutions $\varphi_G^\alpha(x,t) \neq \overline{\varphi}$, $\varphi_G^\alpha \in \mathcal{H}_{\overline{\varphi}}$, whose propagation in Ω_R in the time interval $t \in [0,T]$ is that of free waves (**Goldstone-like solutions**).*

Proof.

i) For solutions $\varphi \in \mathcal{H}_{\overline{\varphi}}$, i.e. of the form $\varphi = \overline{\varphi} + \chi$, $\chi \in H^1$, the conservation of the current $j_\mu = (\partial_\mu \varphi) T^\alpha \varphi$, associated with the generator T^α, (without loss of generality we can take φ real and T^α antisymmetric), reads

$$0 = \partial_\mu j^\mu = \Box \chi_i\, T_{ij}^\alpha\, \varphi_j = \Box(\chi_i\, T_{ij}^\alpha\, \overline{\varphi}) + \Box \chi_i\, T_{ij}^\alpha\, \chi_j, \qquad (9.4)$$

and by the invariance of the potential, the second term can be written as $U'(\overline{\varphi}+\chi)_i\, T_{ij}^\alpha\, \overline{\varphi}_j$. (In the quantum case, thanks to the vacuum expectation value, one has only the analogue of the first term and the proof gets simpler).

Now, for small amplitude solutions χ, the asymptotic limits are governed by the behaviour of $\mathcal{U}'(\chi) \equiv U'(\overline{\varphi} + \chi)$ near $\chi = 0$ and in this region, by the invariance of the potential, one has

$$\mathcal{U}_i'(\chi)\, (T^\alpha\, \overline{\varphi})_i = U_{ijk}'''(\overline{\varphi})\, \chi_j \chi_k\, (T^\alpha\, \overline{\varphi})_i + O(\chi^3).$$

This implies that the small amplitude mode $\chi^\alpha \equiv \chi_i\, (T^\alpha\, \overline{\varphi})_i$ satisfies a nonlinear wave equation with an effective potential which vanishes to a degree $p \geq 3$ near $\chi = 0$. Thus, the large time decay of the nonlinear term appearing in the corresponding Yang-Feldman equation is not worse than in the case of a wave equation with potential vanishing with degree $p \geq 3$

near the origin (other massive modes occurring in $\mathcal{U}'$ have faster decay properties). Then, one can appeal to standard results[43] to obtain the existence of the asymptotic limits $\chi^{\alpha}_{\pm}(t)$ satisfying the free wave equation.

ii) The existence of free waves $\varphi(x, t) = \overline{\varphi} + \chi(x, t)$ within a given region Ω_R in the time interval $[0, T]$ is equivalent to $U'(\overline{\varphi} + \chi(x, t)) = 0$, $\forall x \in \Omega_R$, $t \in [0, T]$, so that if the absolute minima of the potential consist of a single orbit $\overline{\varphi} + \chi(x, t) = \exp\left(h^{\alpha}(x, t) T^{\alpha}\right) \overline{\varphi}$, $h^{\alpha}(x, t)$ real $\in H^1$ and for solutions associated to a given generator T^{α}, with $T^{\alpha} \overline{\varphi} \neq 0$, one has solutions of the form

$$\varphi^{\alpha}(x, t) = e^{h(x, t) T^{\alpha}} \overline{\varphi}.$$

Now, the wave equation $\Box \varphi(x, t) = 0$ requires

$$\Box h(x, t) = 0, \quad (\partial_{\mu} h \, \partial^{\mu} h)(x, t) = 0, \tag{9.5}$$

(since T^{α} and $(T^{\alpha})^2$ have different symmetry properties).

This implies that any C^2 function of h also satisfies (9.5) and in particular $\chi^{(\alpha)} \equiv \varphi^{(\alpha)} - \overline{\varphi}$ also does.

Equations (9.5) have solutions of the form $h_k(x, t) = h(k_0 t - \mathbf{k} \cdot \mathbf{x})$, with h an arbitrary C^2 function and $k = (k_0, \mathbf{k})$ a light-like four vector, but they are not in $H^1(\mathbb{R}^s)$ for $s \geq 2$. One can argue more generally that the above equations do not have solutions $h \in H^1(\mathbb{R}^s)$ for $s \geq 2$. In fact, the wave equation requires that the support of the $s + 1$–dimensional Fourier transform $\hat{h}(k)$, $k \in \mathbb{R}^{s+1}$ is contained in $\{k^2 = 0\}$, and the second equation becomes

$$k^2 \int d^{s+1}q \, \overline{\hat{h}}(q - k) \, \hat{h}(q) = 0,$$

since $kq - q^2 = k^2 - (k - q)^2$, $(k - q)^2 \hat{h}(k - q) = 0$. Thus,

$$H(k) \equiv \int d^{s+1}q \, \overline{\hat{h}}(q - k) \, \hat{h}(q)$$

must have support in $k^2 = 0$. Now, the sum of two light-like four vectors $k - q$, q may be a light-like vector k only if $\mathbf{k}$ and $\mathbf{q}$ are parallel or antiparallel, corresponding to sign $k_0 q_0 = +1$ or $= -1$, respectively, i.e. only if $q = \lambda k$, $\lambda \in R$. Hence, if $k \in$ supp H and q and $q - k$ belong to the support of $\hat{h}$, q must lie in the intersection of the light cone $q^2 = 0$ and the hyperplane $kq = 0$; thus, writing

$$\hat{h}(q) = \delta(q^2) \, h_r(\mathbf{q}), \quad H(k) = \delta(k^2) \, H_r(\mathbf{k}),$$

where δ denotes the Dirac delta function, one has

$$H_r(\mathbf{k}) = \mu(I_k) \int d\lambda \, h_r(\mathbf{k}(1 - \lambda)) \, h_r(\lambda \mathbf{k}),$$

---

[43] H. Pecher. Math. Zeit. **185**, 261 (1984); **198**, 277 (1988).

where μ is the Lebesgue measure and

$$I_k \equiv \{\mathbf{q}; \mathbf{q} \in \mathrm{supp}\, h_r \cap \{kq = 0, k^2 = 0, q^2 = 0\}\}$$

For $s \geq 2$ this appears to exclude that $h \in H^1(\mathbb{R}^s)$.

The above argument indicates that the solutions with the properties of ii) can be constructed, e.g. as

$$\varphi_G^\alpha(x,t) = e^{h_k(x,t)\, f_{R+2T}(\mathbf{x})\, T^\alpha}\, \overline{\varphi},$$

with $f_R(\mathbf{x}) = 1$ for $|\mathbf{x}| \leq R$ and $= 0$ for $|\mathbf{x}| \geq R(1+\varepsilon)$.

The above discussion also shows that in one space dimension $s = 1$ one may find solutions of (9.5) belonging to H^1 and therefore one proves the existence of genuine Goldstone modes all over the space. In fact, any function $h(x - t)$ or $h(x + t)$, $h \in H^1(\mathbb{R})$, is a solution of (9.5).

10 Appendix

A Properties of the Free Wave Propagator

a) $W(t)$ *maps* $\mathcal{S} \times \mathcal{S}$ *into* $\mathcal{S} \times \mathcal{S}$

If $u \in \mathcal{S}(\mathbb{R}^s) \times \mathcal{S}(\mathbb{R}^s)$ ($\mathcal{S}(\mathbb{R}^s)$ is the Schwartz space of C^∞ test functions decreasing at infinity faster than any inverse polynomial), then the solution of the free wave equation is easily obtained by Fourier transform and one has

$$W(t) \begin{pmatrix} \varphi_0(k) \\ \psi_0(k) \end{pmatrix} = \begin{pmatrix} \cos|k|t & (\sin|k|t)/|k| \\ -|k|\sin|k|t & \cos|k|t \end{pmatrix} \begin{pmatrix} \varphi_0(k) \\ \psi_0(k) \end{pmatrix}. \tag{A.1}$$

$\cos|k|t, (\sin|k|t)/|k|$ etc. are multipliers of $\mathcal{S} \equiv \mathcal{S}(\mathbb{R}^s)$ continuous in t and

$$\frac{d}{dt}W(t)\Big|_{t=0} = \begin{pmatrix} 0 & 1 \\ |k|^2 & 0 \end{pmatrix} = K. \tag{A.2}$$

The group property is easily checked.

b) *Hyperbolic character of* $W(t)$. *Huygens' principle.*

Let Ω_{R-t} be concentric spheres in $\mathbb{R}^s$ of radius $R - t$, $0 \leq t \leq R - \delta$, $\delta > 0$, for simplicity centered at the origin, then

$$\|W(t)\,u_0\|_{\Omega_{R-t}} \leq e^{|t|/2}\,\|u_0\|_{\Omega_R}. \tag{A.3}$$

This is a mathematical formulation of Huygens' principle: the norm of $u(t)$ in Ω_{R-t} depends only on the norm of $u(0)$ in Ω_R (influence domain). We start by proving (A.3) for $u \in \mathcal{S} \times \mathcal{S}$. The free wave implies

$$\tfrac{1}{2}\frac{d}{dt}[(\nabla\varphi)^2 + \psi^2] - \nabla \cdot (\psi\nabla\varphi) = 0 \tag{A.4}$$

(energy-momentum conservation) and, since $\varphi\psi = d(\tfrac{1}{2}\varphi^2)/dt$, one has

$$\tfrac{1}{2}\frac{d}{dt}[(\nabla\varphi)^2 + \varphi^2 + \psi^2] - \nabla(\psi\nabla\varphi) = \varphi\psi. \tag{A.5}$$

Now, we integrate the above equation over the cut cone with lower base Ω_R and upper base Ω_{R-t}, and we use Gauss' theorem to transform the volume

F. Strocchi: *Appendix*, Lect. Notes Phys. **732**, 51–60 (2008)
DOI 10.1007/978-3-540-73593-9_10   © Springer-Verlag Berlin Heidelberg 2008

integral into a surface integral. We get

$$\|u(t)\|^2_{\Omega_{R-t}} - \|u(0)\|^2_{\Omega_R} + \int_S dS \; \{\tfrac{1}{2}[(\nabla\varphi)^2 + \varphi^2 + \psi^2]n_0 - \boldsymbol{n} \cdot (\psi\boldsymbol{\nabla}\varphi)\}$$

$$= \int_0^t d\tau \int_{\Omega_{R-\tau}} \varphi(x,\tau)\psi(x,\tau)\, d^s x, \qquad (A.6)$$

where S is the three-dimensional surface defined by $|x| = R - \tau$, $0 \le \tau \le t$ and $n = (\boldsymbol{n}, n_0)$ is its outer normal. Since $n_0 > 0$ and $|\boldsymbol{n}| = n_0$, we have that the function in curly brackets in the left hand side of (A.6) is greater than

$$n_0 \tfrac{1}{2}[(\nabla\varphi)^2 + \varphi^2 + \psi^2 - 2|\psi|\,|\nabla\varphi|] \ge 0.$$

Furthermore, by the inequality $a^2 + b^2 \ge 2ab$, the integral over $\Omega_{R-\tau}$ on the r.h.s. of (A. 6), is majorized by $\|u(\tau)\|^2_{\Omega_{R-\tau}}$. Hence we get

$$\|u(t)\|^2_{\Omega_{R-t}} \le \|u(0)\|^2_{\Omega_R} + \int_0^t d\tau \, \|u(\tau)\|^2_{\Omega_{R-\tau}}. \qquad (A.7)$$

Now, by *Gronwall's lemma*[44], if a non-negative continuous function $F(t)$ satisfies

$$F(t) \le A(t) + \int_0^t d\tau B(\tau) F(\tau), \qquad (A.8)$$

with $A(t), B(t)$ both continuous and non-negative and $A(t)$ non-decreasing, then

$$F(t) \le A(t) \exp(\int_0^t B(\tau)d\tau). \qquad (A.9)$$

By applying Gronwall's lemma to (A.7) one obtains (A.3) and the hyperbolic character of $W(t)$ on $\mathcal{S} \times \mathcal{S}$.

Equation (A.3) also implies that $W(t)$ is a continuous operator with respect to the X_{loc} topology and then it can be extended from the dense domain $\mathcal{S} \times \mathcal{S}$ to whole X_{loc} preserving (A.3) and the group law.

c) $W(t)$ *is a strongly continuous group*
Thanks to the group law, it is enough to show the strong continuity at $t = 0$, namely that, for any bounded region V in $\mathbb{R}^S$,

$$\lim_{t \to 0} \|(W(t) - 1)u\|_V = 0, \qquad \forall u \in X_{loc}. \qquad (A.10)$$

Equation (A.10) is obvious for $u \in \mathcal{S} \times \mathcal{S}$ (see (A.1)) and it can be extended to the whole X_{loc} by using (A.3). In fact, if $u_j \in \mathcal{S} \times \mathcal{S}$ and $u_j \to u$

---

[44] See e.g. G. Sansone and R. Conti, *Non-linear Differential Equations*, Pergamon Press 1964, p. 11.

in X_{loc}, one has

$$\|(W(t)-1)u\|_V \leq \|(W(t)-1)u_j\|_V + \|(W(t)-1)(u_j-u)\|_V$$

By (A.3), the latter term is majorized by $(e^{\frac{1}{2}}+1)\|u_j - u\|_{\Omega_R}$, where Ω_R is a sphere such that $\Omega_{R-1} \supset V$, and therefore can be made arbitrarily small.

One can show that the domain of the generator K is $H^2_{loc}(\mathbb{R}^S) \oplus H^1_{loc}(\mathbb{R}^S)$; in fact, $\forall u \in X_{loc} \subset \mathcal{S}' \times \mathcal{S}'$, in the distributional sense from (A.2) one has

$$K\begin{pmatrix}\varphi \\ \psi\end{pmatrix} = \begin{pmatrix}\psi \\ \Delta\varphi\end{pmatrix}.$$

The condition that the r.h.s. belongs to X_{loc} gives $\psi \in H^1_{loc}(\mathbb{R}^S)$ and $\Delta\varphi \in L^2_{loc}(\mathbb{R}^S)$, which is equivalent to $\varphi \in H^2_{loc}(\mathbb{R}^S)$.

B The Cauchy Problem for Small Times

Theorem B.1. *If $f(u)$ satisfies a local Lipschitz condition, then properties 1), 2), 3), 4), listed in Chap. 4, hold.*

Proof.[45]
1) One has to check that $W(t-s)f(u(s))$ is an integrable function; it is enough to show that it is a continuous function in the X_{loc} topology. To this purpose, we consider the inequality ($f_s \equiv f(u(s))$)

$$\|W(t-s)f_s - W(t-s')f_{s'}\|_{\Omega_{R-t}} \leq \qquad\qquad (B.1)$$
$$\leq \|(W(t-s) - W(t-s'))f_s\|_{\Omega_{R-t}} + \|W(t-s')(f_s - f_{s'})\|_{\Omega_{R-t}}.$$

The first term on the right hand side goes to zero as $s' \to s$ as a consequence of the strong continuity of $W(t)$ on X_{loc} (see Appendix A, c)). The second term can be estimated by using the hyperbolic character of $W(t)$

$$\|W(t)u_0\|_{\Omega_{R-t}} \leq e^{|t|/2}\|u_0\|_{\Omega_R}$$

(see Appendix A, b)) and the local Lipschitz property of f

$$\|W(t-s')(f_{s'} - f_s)\|_{\Omega_{R-t}} \leq Ae^{|t-s'|/2}\|f_{s'} - f_s\|_{\Omega_R} \leq$$
$$\leq Ae^{|t-s'|/2}\|u(s') - u(s)\|_{\Omega_R}.$$

The r.h.s. goes to zero as $s' \to s$ if $u(t)$ is continuous in time.

2), 3) For any two solutions $u_1(t), u_2(t)$, continuous in time, by the hyperbolicity of the free wave equation and the local Lipschitz property one has

---

[45] We essentially follow Ref. I. quoted in footnote 4, to which we refer for a more detailed and general discussion.

$$\|u_1(t) - u_2(t)\|_{\Omega_{R-t}} \le e^{t/2}\{\|u_{10} - u_{20}\|_{\Omega_R} +$$

$$+ \int_0^t e^{-s/2} \|f(u_1(s)) - f(u_2(s))\|_{\Omega_{R-s}} ds\}$$

$$\le e^{t/2}\{\|u_{10} - u_{20}\|_{\Omega_R} + \bar{C}(\Omega_R, \rho) \int_0^t e^{-s/2} \|u_1(s) - u_2(s)\|_{\Omega_{R-s}} ds\},$$

where $0 \le t < R/2$ and

$$\rho = \sup_{0 \le t < R/2} \|u_i(t)\|_{\Omega_{R-t}}, \quad (i = 1, 2).$$

Then, by Gronwall's lemma (see Appendix A, (A.9))

$$\|u_1(t) - u_2(t)\|_{\Omega_{R-t}} \le \exp\left[\left(\tfrac{1}{2} + \bar{C}(\Omega_R, \rho)\right) t\right] \|u_{10} - u_{20}\|_{\Omega_R}, \qquad (B.2)$$

which implies uniqueness and, for $u_2 = 0$, it yields the hyperbolic character.

4) We briefly sketch the idea of the proof. We first consider the case in which u_0 has compact support $\subset \Omega_R$, in which case the proof essentially reduces to a fixed point argument. Given $\rho > 0$, and a fixed u_0 with $\|u_0\|_{\Omega_R} < \rho/2$, we consider the operator S

$$(Su)(t) \equiv W(t)u_0 + \int_0^t W(t-s)f(u(s))\, ds \qquad (B.3)$$

which maps $C^0(X_{loc}, \mathbb{R})$ into itself (see step 1, above). For T small enough, (depending on ρ), S is a contraction on the space

$$E(T, \rho) = \{u \in C^0([0, T], X_{loc}); \operatorname{supp} u(t) \subset \Omega_{R+t}; \sup_{0 < t \le T} \|u(t)\|_{\Omega_{R+t}} \le \rho\},$$

which is complete with respect to the metric

$$d(u, v) = \sup_{0 \le t \le T} \|u(t) - v(t)\|_{\Omega_{R+t+1}}.$$

In fact, by using (B.3) (u_0 fixed), the hyperbolic character of $W(t)$ and the local Lipschitz property of $f(u)$, one has, for $0 \le t < T, T$ small enough,

$$\|(Su)(t) - (Sv)(t)\|_{\Omega_{R+T+1}} \le$$

$$\le e^{t/2} \int_0^t ds\, e^{s/2}\, \bar{C}(\Omega_{R+T+1}, \rho)\, \|u(s) - v(s)\|_{\Omega_{R+T+1}} \le$$

$$\le e^{t/2}\, t\, \bar{C}(\Omega_{R+T+1}, \rho)\, d(u, v)$$

and S maps $E(T, \rho)$ into itself since

$$\|(S\,u)(t)\|_{\Omega_{R+T+1}} \le e^{t/2} \left\{ \tfrac{1}{2}\rho + t\,\bar{C}(\Omega_{R+T+1}, \rho)\,\rho \right\} \le \rho.$$

By Banach theorem on contractions, S has a fixed point which is the required solution in the interval $[0, T)$.

In the case in which u_0 does not have a compact support, we introduce a space cutoff putting

$$u_{0n} \equiv \begin{pmatrix} \chi_n \varphi_0 \\ \chi_n \psi_0 \end{pmatrix}, \quad \chi_n(x) \in C_0^\infty(\mathbb{R}^s),$$

$\chi_n(x) = 1$, if $|x| \le n$, $\chi_n(x) = 0$ if $|x| \ge 2n$. Then, by the previous argument, (4.6) has a solution $u_n(t)$.

Now, for any sphere Ω_{R-t}, by using the local Lipschitz condition and Gronwall's lemma, as in the derivation of (B.2), we get

$$\|u_n(t) - u_m(t)\|_{\Omega_{R-t}} \le \exp[(\tfrac{1}{2} + \bar{C}(\Omega_R, \rho)\,t\,]\|u_{0n} - u_{0m}\|_{\Omega_R}$$

and since u_{0n} converges in X_{loc} to u_0 as $n \to \infty$, also u_n converges in X_{loc} and it converges to the solution of (4.6), with initial data u_0.

C The Global Cauchy Problem

To prove Theorem 4.1 we start by establishing the following *a priori* estimate.

Lemma C.1. *If the potential U is such that the local Lipschitz condition and the lower bound condition are satisfied, then any solution $u \in C^0(X_{loc}, [0, T])$ of (4.6) with $\operatorname{supp}_{0 \le t < T} u(t) \subset \Omega_R$ satisfies*

$$\sup_{0 \le t < T} \|u(t)\|_{\Omega_{R+1}} \equiv L < \infty. \tag{C.1}$$

Proof. The proof exploits the energy conservation

$$\frac{d}{dt}\{\tfrac{1}{2}\int_{\Omega_{R+1}} [(\nabla\varphi(t))^2 + (\psi(t))^2]\,d^s x + \int_{\Omega_{R+1}} U(\varphi(t))\,d^s x\} = 0. \tag{C.2}$$

(The above equation follows from the continuity equation for the energy momentum densities and the fact that there is no momentum flux through the boundary of Ω_{R+1}, since $\operatorname{supp} u(t) \subset \Omega_R$.) In fact, putting

$$K(t) \equiv \frac{1}{2} \int_{\Omega_{R+1}} [(\nabla\varphi(t))^2 + (\varphi(t))^2 + (\psi(t))^2]\,d^s x,$$

one gets from (C.2)

$$K(t) = K(0) + \int_{\Omega_{R+1}} d^s x\,[U(\varphi(0)) - U(\varphi(t))] + \int_0^t dt' \int_{\Omega_{R+1}} \varphi(t')\psi(t')\,d^s x.$$

Now, by using the lower bound condition $(-U(\varphi(t)) \leq \alpha + \beta\varphi(t)^2)$ and the inequality $\varphi\psi \leq \frac{1}{2}(\varphi^2 + \psi^2) \leq (\varphi^2 + \psi^2 + (\nabla\varphi)^2)$, we have

$$K(t) \leq (K(0) + \text{const}) + (2\gamma + 1) \int_0^t dt' K(t').$$

Then, by Gronwall's lemma one has

$$K(t) \leq (K(0) + \text{const})e^{(2\gamma+1)|t|}$$

which implies (C.1).

Now, we can sketch the proof of Theorem 4.1. Any $u(\bar{t})$, $0 \leq \bar{t} < T$, defined by the solution for small times, for initial data of compact support, can be chosen as initial data for the equation

$$u(t) = W(t - \bar{t})\, u(\bar{t}) + \int_{\bar{t}}^t W(t - s)f(u(s))\, ds$$

equivalently for the equation

$$v(\tau) = W(\tau)\, v_0 + \int_0^\tau W(\tau - s)\, f(v(s))\, ds, \qquad \text{(C.3)}$$

where $v_0 \equiv u(\bar{t})$, $v(\tau) \equiv u(\tau + \bar{t})$, and by Lemma C.1 $\|v_0\|_{\Omega_{R+1}} < \rho$, $\rho > 2L$. Hence, the argument given in Appendix B can be applied and existence of solutions for (C.3) can be proved for $0 \leq \tau < T_1$, with T_1 depending *only* on ρ. Since $\bar{t}$ can be chosen as close as we like to T this provides a continuation beyond T.

The existence of solutions for initial data with non-compact support is proved by the same argument as at the end of Appendix B.

D The Non-linear Wave Equation with Driving Term

Theorem D.1. *The equation*

$$\delta(t) = W(t)\delta_0 + L(t) + \int_0^t W(t - s)\, g(\delta(s))\, ds, \qquad \text{(D.1)}$$

with $L(t)$, g, δ defined in the proof of Theorem 5.1, $L(0) = 0$, has a unique solution $\delta(t) \in C^0(X, \mathbb{R})$, $X \equiv H^1(\mathbb{R}^s) \oplus L^2(\mathbb{R}^s)$.

Proof. Uniqueness follows from global Lipschitz continuity by the same argument of Appendix B, (B.2), since the driving term $L(t)$ cancels. As in

Appendix B, existence of solutions for small times follows by a fixed point argument applied to the space

$$E(T, \rho) = \{\delta \in C^0([0, T], X), \sup_{0 \leq t \leq T} \|\delta(t)\|_X < \rho\},$$

since

$$(S\delta)(t) \equiv W(t)\,\delta_0 + L(t) + \int_0^t W(t - s)\, g(\delta(s))\, ds$$

is a contraction on $E(T, \rho)$ for T small enough, $\|\delta_0\|_X < \rho/2$. Finally, the continuation beyond T is obtained as in Appendix C, by exploiting the *a priori* estimate

$$\sup_{0 \leq t < T} \|\delta(t)\|_X \equiv L < \infty,$$

which follows from energy conservation $dE(t)/dt = 0$

$$E(t) = \tfrac{1}{2} \int d^s x [(\nabla \chi(t))^2 + (\zeta(t) + \psi_0)^2] - \int \chi(t)\, h\, d^s x + \int G(\chi(s))\, d^s x$$

$(\chi, \zeta, h, G$ defined in Theorem 5.1$)$. In fact, putting

$$H(t) \equiv E(t) + (\gamma + \tfrac{1}{2}) < \chi(t), \chi(t) > + \tfrac{1}{2} < \psi_0, \psi_0 > + <\omega^{-1}h, \omega^{-1}h> \tag{D.2}$$

where $< .,. >$ denotes the scalar product in L^2, $\omega = (-\Delta)^{\frac{1}{2}}$ and γ is the constant occurring in (5.11), one has

$$H(t) = \tfrac{1}{4} < \omega\chi, \omega\chi > + \tfrac{1}{4} < \zeta, \zeta > + \tfrac{1}{2} < \chi, \chi > + < \psi_0 + \tfrac{1}{2}\zeta, \psi_0 + \tfrac{1}{2}\zeta > +$$

$$+ < \omega^{-1}h - \tfrac{1}{2}\omega\chi, \omega^{-1}h - \tfrac{1}{2}\omega\chi > + \int [G(\chi(s)) + \gamma|\chi|^2]\, d^s x \geq \tfrac{1}{4}\|\delta\|_X \tag{D.3}$$

and

$$< \zeta + \psi_0, \zeta + \psi_0 > \leq 2\{< \psi_0 + \tfrac{1}{2}\zeta, \psi_0 + \tfrac{1}{2}\zeta > + \tfrac{1}{4} < \zeta, \zeta >\} \leq 2H \tag{D.4}$$

Hence,

$$H(t) = H(0) + 2(\gamma + \tfrac{1}{2}) \int_0^t d\tau < \chi(\tau), \zeta(\tau) + \psi_0 > \leq$$

$$\leq H(0) + 2(\gamma + \tfrac{1}{2})2 \int_0^t d\tau H(\tau),$$

so that, by (D.3) and by Gronwall's lemma,

$$\frac{1}{4}\|\delta(t)\|_X \leq H(t) \leq H(0) \exp[4(\gamma + \frac{1}{2})t].$$

E Time Independent Solutions Defining Physical Sectors

We briefly discuss the non-linear elliptic problem associated with the investigation of time independent solutions which define physical sectors (see the first reference in footnote 24). For simplicity, we discuss the case $s \geq 3$. By the discussion of Chap. 6, we have to impose the condition $\nabla \varphi \in L^2(\mathbb{R}^s)$.

Proposition E.1. *Let us consider the non-linear elliptic problem* ($U \in C^2$)

$$\Delta \varphi - U'(\varphi) = 0, \quad \varphi \in H^1_{loc}(\mathbb{R}^s), \quad \nabla \varphi \in L^2(\mathbb{R}^s), \quad s \geq 3; \tag{E.1}$$

then,

i) *the function* $\tilde{\varphi}(r, \omega) \equiv \varphi(x)$, $x = r\omega$, $r > 0$, $\omega \in S^{s-1}$ *(the unit sphere of* $\mathbb{R}^s$*), is continuous in* r *and it has a finite limit* $\tilde{\varphi}(\infty, \omega)$ *as* $r \to \infty$, *for almost all* $\omega \in S^{s-1}$, *and the limit is independent of* ω, *briefly*

$$\lim_{|x| \to \infty} \varphi(x) = \varphi_\infty, \tag{E.2}$$

ii) *(E.1), with boundary condition (E.2), does not have solutions unless* φ_∞ *is a stationary point of the potential*

$$U'(\varphi_\infty) = 0, \tag{E.3}$$

iii) *if* φ_∞ *is an absolute minimum of* U, *then* φ *is the unique solution of (E.1) with* φ_∞ *as boundary value at infinity and* $\varphi = \varphi_\infty$.

Proof.
i) By using a mollifier technique, one reduces the proof of the existence of the limit $\tilde{\varphi}(\infty, \omega)$, for almost all $\omega \in S^{s-1}$, to the estimate

$$|\tilde{\varphi}(r, \omega) - \tilde{\varphi}(r_0, \omega)| \leq \int_{r_0}^r \left| \frac{d}{dr'} \tilde{\varphi}(r', \omega) \right| dr' \leq$$

$$\leq \left(\int_{r_0}^r \left| \frac{d\varphi}{dr'} \right|^2 (r')^{s-1} dr' \right)^{\frac{1}{2}} \left(\int_{r_0}^r (r')^{1-s} dr' \right)^{\frac{1}{2}} \leq$$

$$\leq \text{const} \left(\int_{r_0}^r \left| \nabla \varphi \frac{x}{r'} \right|^2 (r')^{s-1} dr' \right)^{\frac{1}{2}} |r^{2-s} - r_0^{2-s}|^{\frac{1}{2}}. \tag{E.4}$$

The independence of ω, for almost all ω, follows from the following fact: if φ is locally measurable and $\nabla \varphi \in L^p(\mathbb{R}^s)$, $1 \leq p \leq s$, then there exists a constant A, depending on f, such that

$$\varphi - A \in L^q(\mathbb{R}^s), \quad \frac{1}{q} = \frac{1}{p} - \frac{1}{s}. \tag{E.5}$$

To see this we define

$$\mathcal{H}^L = \{f \in \mathcal{S}'(\mathbb{R}^s), \nabla f \in L^p(\mathbb{R}^s)\}$$

and associate to each element of $\mathcal{H}^L$ the norm

$$\|f\|_{\mathcal{H}^L} = \|\nabla f\|_{L^p}.$$

The so obtained normed space is complete, i.e. if $F_j \in \mathcal{H}^L$ is a Cauchy sequence, then $\nabla_k F_j$ converges to an $F^{(k)} \in L^p$ and since $\nabla_k F^{(j)} - \nabla_j F^{(k)} = 0$ in the sense of distributions, there exists an f such that $F^{(k)} = \nabla_k f$. It is convenient to consider the quotient space $\mathcal{H} = \mathcal{H}^L / \mathcal{H}_0$, where $\mathcal{H}_0 = \{f \in \mathcal{S}'(\mathbb{R}^s), \nabla f = 0\}$. $C_0^\infty(\mathbb{R}^s)$ is weakly dense in $\mathcal{H}$, i.e. if $h \in (\mathcal{H})^*$, the dual space of $\mathcal{H}$, then $h(g) = 0$, $\forall g \in C_0^\infty(\mathbb{R}^s)$, implies $h = 0$; in fact, if h is a continuous linear functional on $\mathcal{H}$

$$|h(g)| \le \mathrm{const} \|\nabla g\|_{L^p}$$

and by the Riesz representation theorem this implies that there exists a $h \in L^q$ such that

$$h(g) = \int h \nabla g d^s x.$$

Hence, $h(g) = 0$, $\forall g \in C_0^\infty(\mathbb{R})$ implies $0 = \int h \nabla g d^s x = -\int \nabla h g d^s x$, i.e. $\nabla h = 0$, i.e. $h = \mathrm{const}$, i.e. $h = 0$ as a functional on $\mathcal{H}$.

Finally, if $f \in \mathcal{H}$, there exists a sequence $\{f_j \in C_0^\infty(\mathbb{R}^s)\}$ with $f_j \to f$ in $\mathcal{H}$; this implies that ∇f_j converges in $L^p(\mathbb{R}^s)$ and, by Sobolev's inequality

$$\|f_j\|_{L^q} \le \mathrm{const} \|\nabla f_j\|_{L^p},$$

$f_j \to \tilde{f}$ in L^q and $\tilde{f}$ belongs to the same equivalence class of f, i.e. $f = \tilde{f} + \mathrm{const}$.

ii) Since $\tilde{\varphi}(r, \omega)$ is continuous in r and $U \in C^2$

$$\lim_{r \to \infty} U'(\tilde{\varphi}(r, \omega)) = U'(\tilde{\varphi}(\infty, \omega)) = U'(\varphi_\infty).$$

Equation (E.1) implies that also $\lim_{r \to \infty} \Delta \tilde{\varphi}(r, \omega)$ exists and it is independent of ω. Furthermore, $\forall f(r) \in \mathcal{D}(\mathbb{R}^+)$, with $\int_0^\infty f(r) dr = 1$

$$U'(\varphi_\infty) = \lim_{r \to \infty} \Delta \tilde{\varphi}(r, \omega) = \lim_{a \to \infty} (\Delta \tilde{\varphi})(r + a, \omega) =$$

$$= \lim_{a \to \infty} \int_0^\infty dr f(r)(\Delta \tilde{\varphi})(r + a, \omega) =$$

$$= \lim_{a \to \infty} \int_0^\infty dr (\Delta f(r)) \tilde{\varphi}(r + a, \omega) =$$

$$= \varphi(\infty) \int_0^\infty dr \Delta f(r) = 0.$$

iii) If φ_∞ is an absolute minimum

$$\tilde{U}(\varphi) \equiv U(\varphi) - U(\varphi_\infty) \geq 0$$

and the solutions of (E.1) are stationary points of the functional

$$H(\varphi) = \int [|\nabla\varphi|^2 + \tilde{U}(\varphi)]d^s x.$$

Now, by putting $\varphi_\lambda(x) \equiv \varphi(\lambda x)$, $\lambda \geq 0$, we get

$$H_\lambda = \int [|\nabla\varphi_\lambda|^2 + \tilde{U}(\varphi_\lambda)]d^s x = \int [\lambda^{-1}|\nabla\varphi|^2 + \lambda^{-3}\tilde{U}(\varphi)]d^s x$$

and

$$\delta^{(\lambda)} H = \delta\lambda \frac{\partial H_\lambda}{\partial \lambda} = -(\delta\lambda)\lambda^{-2} \int [|\nabla\varphi|^2 + 3\lambda^{-2}\tilde{U}(\varphi)]d^s x.$$

Hence, the condition of stationarity and the positivity of $\tilde{U}$ yield

$$\int |\nabla\varphi|^2 d^s x = 0, \quad \text{i.e.} \quad \nabla\varphi = 0, \quad \text{i.e.} \quad \varphi = \varphi_\infty,$$

and $\tilde{U}(\varphi_\infty) = 0$.

Part II

Symmetry Breaking in Quantum Systems

Introduction to Part II

These notes arose from courses given at the International School for Advanced Studies (Trieste) and at the Scuola Normale Superiore (Pisa) in various years, with the purpose of discussing the structural features and collective effects which distinguish the quantum mechanics of systems with infinite degrees of freedom from ordinary quantum mechanics.

The motivations for considering *systems with infinite degrees of freedom* are many. Historically, the first and one of the most important ones came from the problem of describing particle interactions consistently with the principles of special relativity. As it is well known, the concept of force as "action at a distance" between particles involves the concept of simultaneity and it does not fit into the framework of relativity, unless one is ready to accept highly non-local actions. This is the reason why so far special relativity has provided a beautiful kinematics but no relativistically invariant theory of (classical) particle (action at a distance) interactions. The transmission of energy and momentum by local (or contact) actions leads to the concept of "medium" or *field* as the carrier of the transmitted energy and momentum and therefore to a system with infinite degrees of freedom.

Another important class of physical phenomena, whose description involves infinite degrees of freedom, are those related to the bulk properties of matter. In fact, the intensive properties of systems consisting of a large number $N \sim 10^{27}$ of constituents are largely independent of N and of the occupied volume V, for given fixed density $n = N/V$; therefore, their description greatly simplifies by taking the so-called thermodynamical limit $N \to \infty$, $V \to \infty$ with n fixed. In this way one passes to the limit of infinite degrees of freedom. Collective phenomena, phase transitions, thermodynamical properties etc. could hardly have a simple treatment without such a limit.

The quantization of systems with infinite degrees of freedom started being investigated soon after the birth of quantum mechanics and it was soon realized that new theoretical structures were involved. In particular, the states of an infinite system cannot be described by a single wave function (of an infinite number of variables) as in ordinary quantum mechanics, i.e. the standard Schroedinger representation is not possible. The changes involved were regarded so substantial to deserve the name of second quantization. As emphasized by Segal, Haag, Kastler and others, it is more convenient, logically more economical and actually more general to formulate the principles of quantum mechanics in terms of (the algebra of) observables and states as positive linear functionals or expectations on the observables. This covers both the case of finite degrees of freedom, where the Von Neumann theorem selects the Schroedinger representation in an essentially unique way, and the case of infinite degrees of freedom, for which even the Fock representation is generically forbidden (apart from the free field case).

These notes focus the attention on the mechanism of spontaneous symmetry breaking (SSB). It seems fair to say that the realization of such a possibility represented a real breakthrough in the development of theoretical physics. In fact it is at the basis of most of the recent achievements in Many Body Theory and in Elementary Particle Theory.

In spite of the cheap explanations, the phenomenon of SSB is deep and subtle and crucially involves the occurrence of infinite degrees of freedom. From elementary quantum mechanics, one learns that the symmetries of the Hamiltonian are symmetries of the physical description of the system, which does not mean that the ground state is symmetric, but rather that the symmetry transformations commute with the time evolution. Thus, whenever the symmetry can be implemented by a (physically realizable) correspondence between the states of the systems, no symmetry breaking can be observed.

The way out of this obstruction is the realization that for infinitely extended systems, the algebra of observables, which define a given system, and its time evolution do not select a unique realization of the system, but rather one has more than one "physical world" or (infinite volume thermodynamical) "phase", which are physically disjoints in the sense that no physically realizable operation can lead from one to the other. Technically this corresponds to the existence of inequivalent representations of the algebra of observables. The occurrence of spontaneous symmetry breaking in a given world is then related to its instability with respect to the symmetry transformations. Thus, the lack of symmetry is due to the impossibility of comparing the properties of a state with those of its transformed one, since the latter belongs to a physically disjoint world. The necessary localization in space (and time) of any physically realizable operation and the infinite extension of the system are crucial ingredients for such a phenomenon.

The occurrence of inequivalent representations of the algebra of canonical variables or more generally of observables, for systems with infinite degrees of freedom (briefly infinite systems), is briefly reviewed in Chaps. 1–3.

A general formulation of quantum mechanics of infinitely extended systems is made possible by exploiting the localization properties of the algebra of canonical variables or of observables. As emphasized by Haag, the local structure is the key property and together with the related asymptotic abelianess and cluster property plays a crucial role for the identification of the physically relevant representations and of the "pure" phases. A clear discussion of spontaneous symmetry breaking could not be done without the realization of these points (Chaps. 4–7).

General criteria and non-perturbative constructive approaches to spontaneous symmetry breaking are briefly discussed in Chaps. 8–10 and applied to simple examples in Chaps. 11 and 13. In particular the Ising model displays the discrepancy between the non-perturbative (Ruelle and Bogoliubov) approaches and the perturbative (Goldstone) criterium.

The modification of the general structure for systems at non-zero temperature and the basic role of the KMS condition is reviewed in Chap. 12 and applied to simple examples of many body systems and of quantum fields.

The spontaneous breaking of continuous symmetries and the implication on the energy spectrum are discussed in detail in Chap. 15. The Goldstone theorem is carefully discussed with a critical analysis of its hypotheses. In particular, the integrability of the charge density commutators and the localization properties of the dynamics are argued to be the relevant ingredients for a clear and mathematical control of the Goldstone theorem for non-relativistic systems. The relation between the range of the potential and the critical delocalization of the dynamics leading to an evasion of the Goldstone theorem is worked out in detail beyond the Swieca conjecture. By using a perturbative expansion in time, the critical decay of the potential for the absence of massless Goldstone bosons and the occurrence of an energy gap turns out to be that of the Coulomb potential rather than the one power faster decay predicted by Swieca condition. Such an analysis clarifies the link between spontaneous symmetry breaking in non-relativistic Coulomb systems and in (positive) gauge theories (*Higgs phenomenon and $U(1)$ problem in QCD*); in particular it explains the occurrence of *"massive"* Goldstone bosons associated to symmetry breaking as a consequence of a Coulomb-like delocalization induced by the dynamics in both cases.

The non-zero temperature version of the Goldstone theorem is discussed in Chap. 16, with a careful handling of the distributional problems of the zero momentum limit, which actually gives rise to derivatives of the Dirac delta function. The extension of the Goldstone theorem to the more general case in which the Hamiltonian and the generators of the symmetry group generate a Lie algebra (*non-symmetric Hamiltonians*) provides non-perturbative information on the energy gap of the modified Goldstone spectrum (Chap. 18).

A version of the Goldstone theorem for gauge symmetries in local gauge theories, which accounts for the absence of physical Goldstone bosons (Higgs mechanism), is presented in Chap. 19, by exploiting Gauss' law and an extension of the Goldstone theorem for relativistic local fields, which does not use positivity.

In conclusion the aim of these lectures is to provide an introduction to the quantum mechanics of infinitely extended systems and to the fascinating and important subject of spontaneous symmetry breaking. No pretension of completeness is made about the subject, which has a vast physical and mathematical literature. Notwithstone, the basic mechanism of spontaneous symmetry breaking, apart from the popular accounts which do not convey the relevant mathematical structures, does not seem to be part of the common education of theoretical physics students.

The background knowledge required is reduced to the basic elements of the theory of Hilbert space operators and to the foundations of ordinary quantum mechanics. The presentation does not indulge in the mathematical details, while respecting the mathematical correctness of the arguments, hoping to keep the message clear and direct for a wide audience possibly including mathematical students. The basic ideas and structures are discussed in a way which should be easily implementable with full rigor according to the taste of the reader.

The chapters marked with a * can be skipped in a first reading.

The material presented in these lectures is largely based on collaborations and illuminating discussions with Gianni Morchio, to whom I am greatly indebted.

1 Quantum Mechanics.
Algebraic Structure and States

We briefly review the basic structure of Quantum Mechanics (QM) with the aim of covering both the case of systems with a finite number of degrees of freedom (ordinary QM) as well as the case of systems with an infinite number of degrees of freedom (briefly *infinite systems*).[46]

For this purpose, it is useful to recall that in the original formulation of QM the emphasis has been on the canonical structure, in terms of the canonical variables q, p, in analogy with the classical case. The quantization conditions, which mark the basic difference between classical and quantum mechanics, amounts to replace the Poisson brackets structure, for N degrees of freedom, with the canonical commutation relations (CCR)

$$[\, q_i, \, p_j \,] = i \, \delta_{ij}, \quad [\, q_i, \, q_j \,] = 0 = [\, p_i, \, p_j \,], \quad i, j = 1, 2, ..., N, \qquad (1.1)$$

where δ_{ij} denotes the Kronecker symbol and for simplicity units have been chosen such that $\hbar = 1$. In this way, the abelian algebra of the classical canonical variables is turned into the non-abelian *Heisenberg algebra* $\mathcal{A}_H$.[47]

The CCR imply that the canonical variables q, p cannot both be represented by self-adjoint *bounded* operators in a Hilbert space.[48] This is the source of technical mathematical problems (domain questions etc.), so that it is more convenient to use as basic variables the so-called *Weyl operators*

$$U(\alpha) \equiv e^{i\alpha q}, \quad V(\beta) \equiv e^{i\beta p}, \quad \alpha q \equiv \sum_i \alpha_i \, q_i, \quad \beta p \equiv \sum_i \beta_i \, p_i, \quad \alpha_i, \, \beta_i \in \mathbf{R}$$

and the algebra $\mathcal{A}_W$ generated by them, briefly called the *Weyl algebra*, instead of the Heisenberg algebra.

---

[46] For an elementary introduction to the quantum mechanics of infinite systems, see e.g. F. Strocchi, *Elements of Quantum Mechanics of Infinite Systems*, World Scientific 1985, hereafter referred to as [S 85].

[47] W. Heisenberg, *The Physical Principles of the Quantum Theory*, Dover 1930; P.A.M. Dirac, *The Principles of Quantum Mechanics*, Oxford University Press 1986.

[48] In fact, the CCR imply $i \, n \, q^{n-1} = q^n p - p \, q^n$ and by taking the norms one has $n||q^{n-1}|| \leq 2||q^{n-1}|| \, ||q|| \, ||p||$, i.e. $||q|| \, ||p|| \geq n/2$.

F. Strocchi: *Quantum Mechanics. Algebraic Structure and States*, Lect. Notes Phys. **732**, 67–71 (2008)
DOI 10.1007/978-3-540-73593-9_1 © Springer-Verlag Berlin Heidelberg 2008

The Heisenberg algebra can in any case be recovered under the general regularity condition of strong continuity of $U(\alpha), V(\beta)$, thanks to Stone's theorem.[49]

In terms of the Weyl operators, the Heisenberg commutation relations read

$$U(\alpha)U(\alpha') = U(\alpha + \alpha'), \quad V(\beta)V(\beta') = V(\beta + \beta'),$$

$$U(\alpha)V(\beta) = e^{-i\alpha\beta} V(\beta) U(\alpha). \tag{1.2}$$

The self-adjointness condition on the q, p naturally defines an antilinear $*$ operation in $\mathcal{A}_W$

$$U(\alpha)^* \equiv U(-\alpha), \quad V(\beta)^* \equiv V(-\beta), \tag{1.3}$$

which turns $\mathcal{A}_W$ into a $*$-algebra.

Furthermore, in order to construct more general functions of the canonical variables (than the Weyl exponentials), a criterium of convergence or a topology is needed; for general mathematical and technical reasons, it is convenient to assign a norm $|| \; ||$ to the elements of $\mathcal{A}_W$, with the property

$$||A^*A|| = ||A||^2, \quad \forall A \in \mathcal{A}_W, \tag{1.4}$$

and to consider the norm closure of $\mathcal{A}_W$, still denoted by the same symbol.

It is a general mathematical result that for the Weyl algebra this can be done in one and only one way.[50] A norm with the above property is called a C^*-norm and in this way the Weyl algebra becomes a C^*-algebra. From a physical point of view, the intrinsic meaning of the norm of an element A is that of the maximum absolute value which can be taken by the expectations of A on *any* state. The topology induced by the norm is usually called the uniform (or norm) topology; it is the strongest one and also the one with an intrinsic algebraic meaning.

The above discussion emphasizes the algebraic structure at the basis of quantum mechanics, with the algebra $\mathcal{A}_W$ of canonical variables playing the same kinematical role as the (algebra of the) classical canonical variables. The identification of such an algebra is a preliminary step for the description of a given system and actually can be taken as the basic point for the definition of the system.

More generally, for systems with infinite degrees of freedom and especially for relativistic quantum systems (where the canonical formalism is problematic, see e.g. [S 85]), it is more convenient to identify the algebraic structure, which underlies the definition of the system, with the algebra (with identity) generated by the physical quantities, briefly called *observables*, which can be

---

[49] See e.g. M. Reed and B. Simon, *Methods of Modern Mathematical Physics*, Vol. I, Academic Press 1972, Sect. VIII.4.

[50] J. Slawny, Comm. Math. Phys. **24**, 151 (1972); J. Manuceau, M. Sirugue, D. Testard and A. Verbeure, Comm. Math. Phys. **32**, 231 (1973).

measured on the given system.[51] From an operational point of view, a system is actually defined by its algebra of observables $\mathcal{A}$ and, by appealing to the operational properties of measurements, one can argue that $\mathcal{A}$ has an identity and can be given the structure of C^*-algebra.[52] In the sequel, we shall take the point of view that a quantum system is defined by its algebra of observables $\mathcal{A}$, with the understanding that in many concrete cases it can be identified with the algebra of canonical variables.

The explicit link between the algebra $\mathcal{A}$ and the results of measurements is provided by the concept of state. Just as in the classical case, a *state* Ω is characterized by the expectation values of the canonical variables or more generally of the observables: $< A >_\Omega \equiv \Omega(A)$; namely Ω is a functional $\Omega : \mathcal{A} \to \mathbf{C}$ with the property of being linear and positive

$$\Omega(\lambda A + \mu B) = \lambda \Omega(A) + \mu \Omega(B), \quad \Omega(A^* A) \geq 0, \quad \forall A \in \mathcal{A}, \qquad (1.5)$$

and conventionally normalized to one $\Omega(\mathbf{1}) = 1$. It follows that Ω is a continuous functional[53]

$$|\Omega(A)| \leq ||A||, \quad \forall A \in \mathcal{A}. \qquad (1.6)$$

The above general characterization of states does not only cover the standard case of the so-called pure states Ω, represented by vectors Ψ of a Hilbert space $\mathcal{H}$, (the expectation on such states being given by the matrix elements $< A >_\Omega = (\Psi, A\Psi)$, with $(,)$ the scalar product in $\mathcal{H}$), but also the states, briefly called mixed states, whose expectation values are given by normalized density matrices, namely are of the form

$$< A >_\Omega = \mathrm{Tr}(\rho_\Omega A), \quad \rho_\Omega = \sum_i \lambda_i^\Omega P_i, \quad \lambda_i^\Omega \geq 0, \quad \sum_i \lambda_i^\Omega = 1, \qquad (1.7)$$

with P_i one-dimensional projections.

Quite generally a *pure state* on a C^*-algebra is a state which cannot be decomposed as a convex sum

$$\Omega = \lambda \Omega_1 + (1 - \lambda) \Omega_2, \quad 0 < \lambda < 1, \qquad (1.8)$$

of two other states. Otherwise the state is called a *mixed state*.

The above definition of state is particularly useful for the description of systems with infinite degrees of freedom, briefly *infinite systems*, for which

---

[51] This philosophy has been pioneered by I. Segal, R. Haag, D. Kastler, H. Araki etc., see R. Haag, *Local Quantum Physics*, Springer 1996.

[52] For a simple discussion see F. Strocchi, *An Introduction to the Mathematical Structure of Quantum Mechanics*, World Scientific 2005, hereafter referred to as [SNS 96].

[53] For a handy presentation of the algebraic approach to quantum mechanics see [SNS 96]. A general reference for the algebraic approach to QM is O. Bratteli and D.W. Robinson, *Operator Algebras and Quantum Statistical Mechanics*, Vol. 1, 2, Springer 1987, 1996.

the elementary concept of wave function is not available and even the standard Schroedinger representation of the algebra of canonical variables may not be allowed (as we shall see below). Whereas in the case of a quantum system with N degrees of freedom one may choose, e.g., the maximal abelian algebra generated by the N coordinates $q_1, ..., q_N$ and describe the states of the system by wave functions of such variables, for an infinite system one should consider an infinite set of coordinates and it is problematic to define the corresponding wave functions. Moreover, the Von Neumann uniqueness theorem does not apply to infinite systems and there are in general physically relevant representations of the algebra of canonical variables which are not equivalent to the Schroedinger representations (the physical meaning of this problem shall be discussed below). The virtue of the above definition of state is that it applies in general, without involving the concept of wave function.

It is a deep result of Gelfand, Naimark and Segal,[54] also called the *GNS construction*, that the knowledge of a state Ω in the above sense, namely in terms of its expectations on $\mathcal{A}$, uniquely determine (up to isometries) a representation[55] π_Ω of the canonical variables, or more generally of the observables, as operators in a Hilbert space $\mathcal{H}_\Omega$ which contains a reference vector Ψ_Ω, whose matrix elements reproduce the given expectations

$$(\Psi_\Omega, \pi_\Omega(A) \Psi_\Omega) = \Omega(A), \quad \forall A \in \mathcal{A}. \tag{1.9}$$

The idea of the proof is to associate to each element $A \in \mathcal{A}$ a vector Ψ_A, which will have the meaning of a vector obtained by applying A to the "reference" vector $\Psi_1 = \Psi_\Omega$. If such an association is done in a way which preserves the linear structure of $\mathcal{A}$, (i.e. $\Psi_A + \Psi_B = \Psi_{A+B}$), one gets a vector space $\mathcal{D}_A$ isomorphic to $\mathcal{A}$, which is naturally equipped with a non-negative inner product

$$(\Psi_A, \Psi_B) = \Omega(A^*B).$$

The null elements are those corresponding to the set

$$\mathcal{J} = \{A \in \mathcal{A}; \, \Omega(B^*A) = 0, \quad \forall B \in \mathcal{A}\}. \tag{1.10}$$

$\mathcal{J}$ is a left ideal of $\mathcal{A}$, i.e. a linear subspace such that $\mathcal{A}\mathcal{J} \subseteq \mathcal{J}$ and one may consider the quotient $\mathcal{A}/\mathcal{J}$ and correspondingly the equivalence classes of vectors $\Psi_{[A+\mathcal{J}]} = \Psi_{[A]} \in \mathcal{D}_A/\mathcal{D}_\mathcal{J} \equiv \mathcal{D}_\Omega$. In this way, the inner product becomes strictly positive on $\mathcal{D}_\Omega$, which is therefore a pre-Hilbert space, and by completion one gets a Hilbert space $\mathcal{H}_\Omega = \overline{\mathcal{D}_\Omega}$.

---

[54] M.A. Naimark, *Normed Rings*, Noordhoff 1964.

[55] We recall that a representation π of a C^*-algebra in a Hilbert space $\mathcal{H}$ is *-homomorphism π of $\mathcal{A}$ into the C^*-algebra of bounded (linear) operators in $\mathcal{H}$, i.e. a mapping which preserves all the algebraic operations, including the $*$.

The representation is then defined by

$$\pi_\Omega(A)\Psi_{[B]} \equiv \Psi_{[AB]} \tag{1.11}$$

(this equation is well defined since $[B] = [C]$ implies $[AB] = [AC]$). By construction, the vector $\Psi_\Omega \equiv \Psi_{[1]}$ is cyclic with respect to $\pi_\Omega(\mathcal{A})$, namely $\pi_\Omega(\mathcal{A})\Psi_\Omega$ is dense in $\mathcal{H}_\Omega$; moreover $||\pi_\Omega(A)||_{\mathcal{H}_\Omega} \leq ||A||$, thanks to the continuity of Ω.

The so constructed representation is unique up to unitary equivalence. In fact, if π' is another representation in a Hilbert space $\mathcal{H}'$ with a cyclic vector Ψ' such that

$$(\Psi',\, \pi'(A)\,\Psi') = \Omega(A),$$

then the mapping $U : \pi_\Omega(A)\Psi_\Omega \rightarrow \pi'(A)\,\Psi'$ and its inverse U^{-1} are defined on dense sets and preserve the scalar products, so that they are unitary and $\pi'(A) = U\pi(A)U^{-1}$.

The GNS representation π_Ω defined by a state Ω is irreducible iff Ω is pure.

As a relevant application of the above result, we consider a *-automorphism α of $\mathcal{A}$, namely an invertible mapping of $\mathcal{A}$ into $\mathcal{A}$, which preserves all the algebraic operations including the * (also called *algebraic symmetry*). If the state Ω is invariant in the sense that

$$\Omega(\alpha(A)) = \Omega(A), \quad \forall A \in \mathcal{A}, \tag{1.12}$$

then, in the GNS representation defined by Ω, such an automorphism is implemented by a unitary operator U_α with

$$U_\alpha \Psi_\Omega = \Psi_\Omega, \quad U_\alpha \pi_\Omega(A)\Psi_\Omega = \pi_\Omega(\alpha(A))\Psi_\Omega. \tag{1.13}$$

Therefore, all the matrix elements are invariant under the operation which implements α, and briefly one says that α gives rise to a symmetry of the states of the Hilbert space $\mathcal{H}_\Omega$. Thus, the invariance of Ω under α implies that α is a symmetry of the physical world or phase defined by Ω through the GNS construction.

2 Fock Representation

The general lesson from the GNS theorem is that a state Ω on the algebra of observables, namely a set of expectations, defines a realization of the system in terms of a Hilbert space $\mathcal{H}_\Omega$ of states with a reference vector Ψ_Ω which represents Ω as a cyclic vector (so that all the other vectors of $\mathcal{H}_\Omega$ can be obtained by applying the observables to Ψ_Ω). In this sense, a state identifies the family of states related to it by observables, equivalently accessible from it by means of physically realizable operations. Thus, one may say that $\mathcal{H}_\Omega$ describes a closed world, or phase, to which Ω belongs. An interesting physical and mathematical question is how many closed worlds or phases are associated to a quantum system. In the mathematical language this amounts to investigating how many inequivalent (physically acceptable) representations of the observable algebra, which defines the system, exist.

For this purpose we remark that, given a pure state Ω, all the states defined by vectors of the Hilbert space $\mathcal{H}_\Omega$ of the GNS construction define (unitarily) equivalent representations; in fact, the corresponding GNS Hilbert spaces can be identified, and any element $A \in \mathcal{A}$ is represented by the same operator $\pi_\Omega(A)$ in all cases. Also the mixed states defined by density matrices in $\mathcal{H}_\Omega$ define essentially the same representation. In fact, the equation

$$\Omega_\rho(A) \equiv \mathrm{Tr}(\rho\, \pi_\Omega(A)) = \sum_i \lambda_i \left(\Psi_i,\, \pi_\Omega(A)\, \Psi_i\right) = \sum_i \lambda_i\, \Omega_i(A), \qquad (2.1)$$

where $\Psi_i \in \mathcal{H}_\Omega$, expresses Ω_ρ as a convex linear combination of states which define representations equivalent to π_Ω. Technically one says that π_{Ω_ρ} is *quasi equivalent* to π_Ω, meaning that it can be decomposed into a sum of representations equivalent to π_Ω.

The set of states of the form (2.1) is called the *folium* of the representation π_Ω and can be interpreted as the set of the states which are accessible from Ω by observable "operations", i.e. the *closed world* of states associated with Ω.

One may wonder whether the above notions are physically important, since they are not usually brought up in the standard presentations of quantum mechanics. The reason is that, contrary to the infinite dimensional case, for systems with a finite number of degrees of freedom, under very general regularity conditions, there is only one irreducible representation of the Weyl algebra, the so-called Fock representation, all the others being unitarily equivalent to it. According to the above discussion, one may then say that there

F. Strocchi: *Fock Representation*, Lect. Notes Phys. **732**, 73–79 (2008)
DOI 10.1007/978-3-540-73593-9_2   © Springer-Verlag Berlin Heidelberg 2008

is only one folium, or only one closed world, available for the representations of the Weyl algebra.

From a conceptual point of view, such result (Von Neumann theorem) explains why, for systems with a finite number of degrees of freedom, the distinction between the algebraic structure of the canonical variables and the states is not so relevant, since there is only one Hilbert space of states for a given quantum system. In the language of Statistical Mechanics one could say that there is only one phase. On the other hand, for infinite systems the occurrence of inequivalent representations, i.e. of different "phases" or different disjoint worlds, is the generic situation.

In the following, we shall discuss a simplified version of Von Neumann's theorem, which characterizes the Fock representation in terms of the number operator.[56]

We allow for infinite degrees of freedom and we consider regular representations of the corresponding (infinite dimensional) Weyl algebra, namely representations π such that $\pi(U(\alpha))$, $\pi(V(\beta))$ with α, β any finite component vectors, are strongly continuous in α, β. This is the standard regularity assumption underlying the analysis of representations of Lie groups; it appears very general since, for separable spaces, it is equivalent to the condition that the matrix elements of $\pi(U(\alpha))$, $\pi(V(\beta))$ are measurable functions. Furthermore, by Stone's theorem, such a regularity condition is equivalent to the existence of the generators. Thus, we have a representation of the (infinite dimensional) Heisenberg algebra $\mathcal{A}_H$ and we may assume that there is a common dense domain D for $\mathcal{A}_H$. The representation is said to be irreducible if any (bounded) operator which commutes with $\pi(\mathcal{A}_H)$ on D is a multiple of the identity. In the following, the symbol A will be used to denote both an abstract element of $\mathcal{A}_H$ as well its representative in the concrete representation we are considering.

For the following purposes, it is convenient to introduce the so-called annihilation and creation operators

$$a_j \equiv (q_j + i\,p_j)/\sqrt{2}, \quad a_j^* = (q_j - i\,p_j)/\sqrt{2}, \tag{2.2}$$

and the so-called number operator $N_j \equiv a_j^* a_j$. The physical meaning of such operators will be discussed below.

The Heisenberg commutation relations give

$$[\,a_j, a_k^*\,] = \delta_{jk}, \quad [\,a_j, a_k\,] = 0 \tag{2.3}$$

---

[56] For a proof of Von Neumann theorem see e.g. [SNS 96]. For the characterization of the Fock representation in terms of existence of the number operator, see G.F. Dell'Antonio and S. Doplicher, J. Math. Phys. **8**, 663 (1967); J.M. Chaiken, Comm. Math. Phys. **8**, 164 (1967); Ann. Phys. **42**, 23 (1968) and references therein.

and

$$[N_j,\, a_k] = -\delta_{jk}\, a_k. \tag{2.4}$$

Proposition 2.1. *In an irreducible representation of the Heisenberg algebra with domain D, the following properties are equivalent*

1) the total number operator $N = \sum_j N_j$ exists in the sense that $\forall \alpha \in \mathbf{R}$

$$strong - \lim_{K \to \infty} e^{i\,\alpha\, \sum_j^K a_j^* a_j} = e^{i\,\alpha\, N} \equiv T(\alpha), \quad \forall \alpha \in \mathbf{R} \tag{2.5}$$

exists on D and defines a one-parameter group of unitary operators $T(\alpha)$ strongly continuous in α, leaving D stable, so that its generator N exists;
2) there exists a vector Ψ_0, called the Fock (vacuum) vector, such that

$$a_j\, \Psi_0 = 0, \quad \forall j. \tag{2.6}$$

In this case the representation is called a Fock representation.

Proof. Property 1) and the commutation relations imply that

$$T(\alpha)\, a_j\, T(\alpha)^{-1} = e^{-i\alpha}\, a_j$$

and therefore $[T(2\pi), \mathcal{A}_H] = 0$. By the irreducibility of $\mathcal{A}_H$ it follows that $T(2\pi) = 1 \exp i\,\theta$, so that $T'(\alpha) \equiv T(\alpha) \exp\left(-i\,\alpha\,\theta/2\pi\right)$ satisfies $T'(2\pi) = 1$. By using this condition in the spectral representation of $T'(\alpha)$

$$T'(\alpha) = \int_{\sigma(N')} dE(\lambda)\, e^{i\alpha\lambda}, \quad N' \equiv N - \theta/2\pi,$$

where $\sigma(N')$ denotes the spectrum of N', one concludes that the projection valued spectral measure must be supported on a subset of $\mathbf{Z}$, i.e. the spectrum of N' and therefore of N is discrete. Now, if $\lambda > 0$ is a point of the spectrum of N and Ψ_λ a corresponding eigenvector, then

$$0 < \lambda \|\Psi_\lambda\|^2 = (\Psi_\lambda,\, N\,\Psi_\lambda) = \sum_j \|a_j \Psi_\lambda\|^2,$$

so that there must be at least one j such that $a_j \Psi_\lambda \neq 0$ and one has

$$T(\alpha)\, a_j \Psi_\lambda = e^{i(\lambda-1)\alpha}\, a_j\, \Psi_\lambda.$$

Thus, also $\lambda - 1 \in \sigma(N)$ and, since the spectrum of N is non-negative, in order that this process of lowering the eigenvalues terminates, $\lambda = 0$ must be a point of the spectrum of N and

$$a_j\, \Psi_0 = 0, \quad \forall j. \tag{2.7}$$

Conversely, if the Fock vacuum Ψ_0 exists, then $\mathcal{A}_H\, \Psi_0 = \mathcal{P}(a^*)\, \Psi_0$, where $\mathcal{P}(a^*)$ denotes the polynomial algebra generated by the a^*'s. On such a domain, which is dense by the irreducibility of $\mathcal{A}_H$, N exists as a self-adjoint operator and the exponential series converges strongly and defines a one-parameter group of unitary operators, since the monomials of a^* applied to Ψ_0 yield eigenstates of N and generate such a domain.

In the case of a finite number of degrees of freedom, the above argument can be turned into an analog of Von Neumann's theorem by proving that $\forall j$, N_j exists as a self-adjoint operator on D, as a consequence of the regularity condition.[57]

As in the finite dimensional case, all irreducible Fock representations are unitarily equivalent and one can actually speak of one (irreducible) Fock representation. In fact, given any two of them, say π_1, π_2, with Fock vectors Ψ_{01}, Ψ_{02}, respectively, the mapping U defined by

$$U\,\Psi_{01} = \Psi_{02}, \quad U\,\pi_1(A)\,\Psi_{01} = \pi_2(A)\,\Psi_{02}, \quad \forall A \in \mathcal{A}_H$$

and its inverse U^{-1} are defined on dense sets since, by irreducibility, Ψ_{01}, Ψ_{02} are cyclic vectors. Furthermore, since the matrix elements

$$(\pi_i(A)\,\Psi_{0i}, \, \pi_i(B)\,\Psi_{0i}), \quad i = 1, 2, \quad \forall A, B \in \mathcal{A}_H$$

only involve the canonical commutation relations and the Fock condition, (2.6), they are equal, so that U is unitary.

It is worthwhile to remark that, in an irreducible Fock representation, the zero eigenvalue of N has multiplicity one. In fact, if Ψ' is orthogonal to Ψ_0 and satisfies $N\Psi' = 0$, then $a_j\,\Psi' = 0$, $\forall j$ and for any polynomial $P(a^*)$ of the a^* one has

$$(\Psi', P(a^*)\,\Psi_0) = (P(a)\,\Psi', \Psi_0) = 0.$$

This implies $\Psi' = 0$, by the cyclicity of Ψ_0 with respect to the polynomial algebra generated by the a^*.

It is clear from the above argument that the characteristic feature of a Fock representation is that the states of its Hilbert space can be described in terms of the eigenvalues of the N_j, all of which exist as self-adjoint operators on D since they are dominated by N. For this reason the Fock representation is also called the *occupation number representation*.

It should be stressed that only in the Fock representation the annihilation and creation operators a_j, a_j^* have a simple interpretation, namely that of decreasing or increasing the eigenvalues of N_j (or of N). Even in this case, however, their physical meaning may not be transparent, since $N_j = (p_j^2 + q_j^2 - 1)/2$ may not be related to a relevant observable (see e.g. the case of the hydrogen atom or even of the free particle). A special case is that of a system of free harmonic oscillators, where N_j is related to the Hamiltonian for the j-th degree of freedom and a_j, a_j^* respectively annihilate and create elementary excitations of the system. By the same reasons, in general the

---

[57] See e.g. O. Bratteli and D.K. Robinson, *Operator Algebras and Quantum Statistical Mechanics*, Vol. 2, Springer 1996, Sect. 5.2.3; see also [SNS 96].

Fock state is not related to the possible ground state nor does it in general have a simple physical meaning.

The picture emerging from the case of a system of harmonic oscillators however suggests that the occupation number representation may be useful for describing systems whose states can be described in terms of number of elementary excitations. In this case, the index j may be taken to label the j-th excitation (j may denote the set of quantum numbers which identify such excitation) and a_j, a_j^* decrease and increase, respectively, the number of j-th excitations. The states of the system are then analyzed in terms of products of single excitation (or single particle) states. As a consequence, whereas the Hilbert (sub)space $\mathcal{H}_N$ corresponding to a fixed number N of particles or elementary excitations may not have a ground state, the total Hilbert space (the direct sum of the $\mathcal{H}_N$, $N \in \mathbf{N}$) has the Fock state as ground state (since each elementary excitation has positive energy).

The message from the Proposition 2.1 is that the Fock representation is allowed if N is a good quantum number for the description of the relevant states of the system. This is reasonable in the case of a finite number of degrees of freedom and in the case of non-interacting infinite degrees of freedom (with vanishing mean density). As we shall see below, however, in the case of infinite degrees of freedom, the interaction has generically dramatic effects, in the sense that it usually leads to a redefinition of the degrees of the free theory, with the result that the eigenstates of the total Hamiltonian cannot be described in terms of the eigenstates of the free Hamiltonian, so that N is not a well defined quantum number.

In conclusion, the Fock representation for the algebra generated by the a_j, a_j^* is convenient and physically motivated if such annihilation and creation operators are related to the elementary excitations (or normal modes) which diagonalize the total Hamiltonian. In general, the elementary excitations described by the a_j, a_j^* are those which diagonalize the so-called free (or bilinear) part of the Hamiltonian and therefore the interpretation of such annihilation and creation operators is simple only if the states of the system can be analyzed in terms of elementary excitations corresponding to the free part of the Hamiltonian; as we shall discuss below, for interacting relativistic fields or for many body systems with non-zero density, this is never the case and the Fock representation is not allowed.

The relation between the Fock representation and the free Hamiltonian can be made more precise. To this purpose, we consider a system with infinite degrees of freedom and the associated "free" Hamiltonian

$$H_0 = \sum_i \omega_i\, a_i^*\, a_i,$$

where ω_i denotes the energy of the free i-th excitation. We also assume that there is an energy (or mass) gap, i.e.

$$\omega_i \geq m > 0, \quad \forall i.$$

Then, if we look for a representation of the algebra generated by the a_j, a_j^*, such that H_0 is a self adjoint operator (on the common dense domain), the representation is necessarily a Fock representation. In fact, (the series which defines) H_0 dominates (term by term the series which defines) N

$$N \leq (1/m) H_0$$

and therefore the existence of H_0 entails the existence of N.

Example 2.1. Free scalar field. As an example, we briefly review the quantization of a free massive scalar real field. The problem is to find the (operator-valued distributional) solution of the Klein-Gordon equation

$$(\Box + m^2)\varphi(x) = 0,$$

satisfying the equal time canonical commutation relations ($\pi(x) = \dot\varphi(x)$)

$$[\varphi(\mathbf{x},0),\, \pi(\mathbf{y},0)] = i\,\delta(\mathbf{x} - \mathbf{y}),$$

$$[\varphi(\mathbf{x},0),\, \varphi(\mathbf{y},0)] = [\pi(\mathbf{x},0),\, \pi(\mathbf{y},0)] = 0. \tag{2.8}$$

In contrast with the classical case, the canonical relations (2.8) imply that in order to get well defined operators one must (at least) smear the fields with test functions of the space variables, typically $f \in C^\infty(\mathbf{R}^s)$, s = space dimensions and of fast decrease, (briefly $f \in \mathcal{S}(\mathbf{R}^s)$). Thus, from a mathematical point of view, the fields $\varphi(x), \pi(x)$ have to be regarded as operator valued distributions.

The algebra $\mathcal{A}_W$ of canonical variables can be thought of as generated by the exponentials of the real fields $\varphi(f)$, $\pi(g)$, smeared with test functions $f, g \in \mathcal{S}(\mathbf{R}^s)$. Among the many possible representations of such an infinite dimensional Weyl algebra, a selection criterium is that one has a well defined Hamiltonian.

Now, quantization of the classical Hamiltonian

$$H = \tfrac{1}{2} \int d^s x \left[(\nabla \varphi)^2(x) + \pi^2(x) + m^2 \varphi^2(x) \right]$$

requires some care, because the above formal integral involves both the definition of the product of distributions at the same point (ultraviolet (UV) singularities) and the integration over an infinite volume (infrared (IR) singularities). Thus, in contrast with the standard case of finite degrees of freedom, quantization of the classical expressions requires a regularization and/or a renormalization.

In the case of free fields, the UV renormalization of the Hamiltonian is regarded as trivial (the problem is not even mentioned in most text-books). It is obtained by reordering the products of operators, say AB, so that the creation operators stay on the left and the annihilation operators on the right as if they commute; such a procedure is called Wick ordering and denoted by $: AB :$. Then, in a finite volume with periodic boundary conditions the momentum can take only discrete values k_j and one has $(\omega_j \equiv (k_j^2 + m^2)^{1/2})$,

$$H_{ren} = \tfrac{1}{2} \int_V d^s x \, : [(\nabla \varphi)^2(x) + \pi^2(x) + m^2 \varphi^2(x)] := \sum_j \omega_j a_j^* a_j,$$

where $\sqrt{2} \, a_j \equiv \sqrt{\omega_j} \tilde{\varphi}(k_j) + i(\sqrt{\omega_j})^{-1} \tilde{\pi}(k_j)$, and the tilde denotes the Fourier transform. By the above argument, the condition that H_0 be well defined selects the Fock representation.[58] It is not difficult to show that one has a well defined operator also in the infinite volume limit, when the momentum becomes a continuous variable and

$$H_0 = \tfrac{1}{2} \int d^s x \, : [(\nabla \varphi)^2(x) + \pi^2(x) + m^2 \varphi^2(x)] :$$

$$= \int d^s k \, \omega(k) \, a^*(k) \, a(k), \quad \omega(k) = (k^2 + m^2)^{1/2}$$

is well defined on the dense domain obtained by applying the polynomial algebra of the $a^*(f)$ to Ψ_0, since $H_0 \Psi_0 = 0$ and the commutator is well defined $[H_0, a^*(f)] = a^*(\omega f)$.

The massless case, $m = 0$, deserves a comment, since the number operator is no longer dominated by the free Hamiltonian. In fact, in this case there are representations (the so-called non-Fock coherent state representations) in which H_0 is well defined, but N is not.

---

[58] H.J. Borchers, R. Haag and B. Schroer, Nuovo Cim. **29**, 148 (1963). For a general look at free fields from the point of view of representations of the algebra of canonical variables (canonical quantization), see A.S. Wightman and S. Schweber, Phys. Rev. **98**, 812 (1955); S.S. Schweber, *Introduction to Relativistic Quantum Field Theory*, Harper and Row 1961.

3 Non-Fock Representations

As anticipated in the previous discussions, the Fock representation is very special to the finite dimensional case and to free fields. Actually, as a consequence of Proposition 2.1, non-Fock representations are required in order to describe many particle systems with non-zero density in the thermodynamical limit

$$N \to \infty, \quad V \to \infty, \quad N/V \equiv n \neq 0.$$

In fact, in the Fock representation, $\forall \Psi$ in the domain of N, if N_V denotes the (operator) number of particles in the volume V, one has

$$\|n\Psi\| = \lim_{V \to \infty} V^{-1} \|N_V \, \Psi\| \leq \lim_{V \to \infty} V^{-1} \|N \, \Psi\| = 0.$$

Actually, for systems of non-zero density, in the thermodynamical limit, the free Hamiltonian need not be defined even in the free case; only the energy per unit volume is required to be finite.[59]

In the following, we shall present arguments, on the basis of simple examples, which indicate the need of non-Fock representation, also for systems with zero density, in order to get well defined Hamiltonians.

Quite generally, in the case of interacting fields, the definition of the formal Hamiltonian, typically of the form (in a finite volume)

$$H = \sum_i \omega_i \, a_i^* \, a_i + g H_{int}(a, a^*), \tag{3.1}$$

could in principle be easily obtained if one could find the annihilation and creation operators A_i, A_i^*, corresponding to so-called normal modes which diagonalize the Hamiltonian:

$$H = \sum_i E_i A_i^* \, A_i + E_0, \tag{3.2}$$

where E_0 is a constant. In quantum field theory (QFT), such normal mode

---

[59] For the mathematical discussion of the free Bose gas and for the free fermion gas see H. Araki and E.J. Woods, J. Math. Phys. **4**, 637 (1963); H. Araki and W. Wyss, Helv. Phys. Acta **37**, 139 (1964). For a general account, see O. Bratteli and D.W. Robinson, *Operator Algebras and Quantum Statistical Mechanics*, Vol. II, Springer 1996. A simple discussion is given in Sect. 7.2.

F. Strocchi: *Non-Fock Representations*, Lect. Notes Phys. **732**, 81–87 (2008)
DOI 10.1007/978-3-540-73593-9_3   © Springer-Verlag Berlin Heidelberg 2008

operators are the so called asymptotic fields A_i^{as}, A_i^{as*}, $(as = in/out$ related by the S-matrix).[60]

By the argument of the previous section, in the general case of mass gap, the Hamiltonian (3.2) is a well defined operator only if one uses a Fock representation for the A_i^{as}, A_i^{as*}, i.e., in QFT, a representation defined by a Fock vacuum Ψ_0 for the asymptotic fields

$$A_i^{as} \Psi_0 = 0, \quad \forall i.$$

Such a representation is almost never a Fock representation for the original canonical variables, which instead diagonalize the free part H_0 of the Hamiltonian. In this case, in the representation in which H is well defined, H_0 cannot be well defined (only the sum $H_0 + g\,H_{int}$ is so). Thus, from a mathematical point of view, due to the infinite number of degrees of freedom, the interaction is almost never a small perturbation with respect to the free Hamiltonian.

The above arguments against the use of the Fock representation for the canonical variables a, a^*, in terms of which the model is formally defined, can be turned into a theorem (*Haag theorem*). To this purpose, we consider systems described by canonical variables or fields which have localization properties, i.e. which can be written in the form

$$a_i = \psi(f_i) = \int d^s x\, f_i(\mathbf{x})\, \psi(\mathbf{x}), \quad [\psi(\mathbf{x}), \psi^*(\mathbf{y})] = \delta(\mathbf{x} - \mathbf{y}), \qquad (3.3)$$

where $\{f_i\}$ is an orthonormal set of real L^2 regular functions, e.g. $f_i \in \mathcal{S}(\mathbf{R}^s)$.

This is the case of systems described by canonical variables a_i, a_i^* associated to free elementary or single particle excitations described by the quantum number i. Then, if $\{f_i\}$ denote a set of (real orthonormal) single "particle" wave functions, (i.e. f_i describes the free particle or elementary excitation in the i-state), one may introduce the following canonical fields

$$\psi(\mathbf{x}) \equiv \sum_i a_i f_i(\mathbf{x}), \quad \psi^*(\mathbf{x}) \equiv \sum a_i^* f_i(\mathbf{x}) \qquad (3.4)$$

and therefore obtain (3.3).

For the algebra generated by canonical variables of the above form, the space translations α_a are naturally defined by

$$\alpha_a\left(\psi(f_i)\right) = \psi(f_{ia}), \quad f_{ia}(\mathbf{x}) \equiv f_i(\mathbf{x} - \mathbf{a}), \qquad (3.5)$$

formally equivalent to $\alpha_a(\psi(\mathbf{x})) = \psi(\mathbf{x} + \mathbf{a})$. Clearly, the space translations define (a one-parameter group of) *-automorphisms (or algebraic symmetries) of the algebra of canonical variables. In each irreducible Fock representation

---

[60] For the general (and rigorous) theory, see R. Jost, *The General Theory of Quantized Fields*, Am. Math. Soc. 1965. Unfortunately, the knowledge of the asymptotic fields is essentially equivalent to the control of the full solution.

the Fock state is the unique translationally invariant state[61] and the space translations are implemented by (strongly continuous) unitary operators $U(a)$

$$U(a)\psi(\mathbf{x})U(-a) = \psi(\mathbf{x} + \mathbf{a}). \tag{3.6}$$

We briefly sketch Haag's theorem[62]. We consider a system described by canonical variables of the form (3.3) and by a Hamiltonian of the form (3.1), with $g \neq 0$, invariant under space translations. We denote by π_g an irreducible representation of the algebra of canonical variables in which H is well defined and it has a translationally invariant ground state Ψ_{0g}. Then, by the argument following the GNS theorem, the space translations are implemented by (a one-parameter group of) unitary operators $U_g(a)$ leaving the ground state invariant. If π_g is a Fock representation for the a_i, a_i^*, then there exists a Fock vacuum Ψ_0 and $U_g(a) = U_{g=0}(a) \equiv U(a)$. In fact $U_g U^{-1}$ commutes with the a_i, a_i^* and by irreducibility it must be a multiple of the identity $\exp(i\theta_g(a))$; moreover, the group law and the continuity in a implies that $\theta_g(a) = \theta_g a$ and therefore by a trivial redefinition of $U_g(a)$ one may get $U_g = U$. This implies that $\Psi_{0g} = \Psi_0$, i.e. the ground state is independent of the coupling constant. It is intuitively clear that it can hardly be so and actually for relativistic systems the above coincidence of the interacting and the free ground states is compatible only with a free theory.[63]

The implications of Haag's theorem, about the impossibility of using the Fock representation for defining the Hamiltonian in the presence of interaction, are rather strong. The standard Rayleigh-Schroedinger perturbative expansion in terms of eigenstates of the free Hamiltonian H_0 requires that H_0 be well defined and this is not possible if the representation in which the total Hamiltonian is well defined is non-Fock. In particular, from a mathematical point of view, Haag's theorem excludes the existence of the so-called interaction picture representation, which is at the basis of the standard expansion in quantum field theory and in many body theory. In fact, the existence of the interaction picture is equivalent to the statement that the representation of the field operators at time t is unitarily equivalent to the representation of free fields and, by the Borchers-Haag-Schroer result discussed above, the latter require a Fock representation; in conclusion, at each time t the representation of the interacting field operators should be equivalent to a Fock representation, contrary to Haag's theorem.

---

[61] In fact, since the number operator commutes with the space translations, the existence of space translationally invariant states can be discussed in each eigenspace of N, say $\mathcal{H}_K$, corresponding to the eigenvalue K, whose vectors are $L^2(\mathbf{R}^{sK})$ functions of sK variables. The space translation invariance would require that such a function does not depend on the sum of the variables, incompatibly with being in L^2.

[62] R. Haag, *On quantum field theories*, Dan. Mat. Fys. Medd. **29** no 12 (1955); *Local Quantum Physics*, Springer 1996.

[63] R.F. Streater and A.S. Wightman, *PCT, Spin and Statistics and All That*, Benjamin-Cummings 1980.

Thus, the solution of the dynamical problem for infinite systems is much more difficult (especially from a mathematical point of view) than in the finite dimensional case, where it is essentially controlled by Kato's theorems[64]. In the infinite dimensional case, one faces the puzzling situation that in order to give a meaning to the Hamiltonian, as an operator in a Hilbert space $\mathcal{H}$, one must specify the representation of the operators a, a^* (or of the fields at time zero), in terms of which H is formally defined. On the other hand, the representation of the a, a^*, in which the dynamics is well defined, is in general non-Fock and its determination involves a non-perturbative control on the theory. This looks as a blind alley. A possible way out of these conceptual difficulties (and also a possible way to recover some of the results of the perturbative expansion) is provided by the constructive strategy[65]. As already mentioned above, the strategy is to regularize the theory by introducing UV and IR cutoffs and to determine the (cutoff-dependent) counter terms needed to get a renormalized Hamiltonian, so that the corresponding (ground state) correlation functions have a reasonable limit when the cutoffs are removed. This is the content of the so-called non-perturbative renormalization, which has been successfully carried out in quantum field theory models in low space-time dimensions ($d = 1 + 1$, $d = 2 + 1$).[66] A simple model, in which such a non-perturbative renormalization can be instructively checked to work, and which also displays the occurrence of non-Fock representations, is the so-called Yukawa model of pion-(heavy)nucleon interaction[67].

To give at least the flavour of how non-Fock representations arise, we list a few simple examples.

Example 3.1. Quantum field interacting with a classical source. We consider a quantum scalar field (see Example 2.1) interacting with a classical (time independent) real source $j(x)$

$$(\Box + m^2)\varphi(x) = gj(x), \tag{3.7}$$

where φ satisfies the equal time canonical commutation relations (2.8). The formal Hamiltonian is ($\omega(k) \equiv (\mathbf{k}^2 + m^2)^{1/2}$)

$$H = \int dk\,\omega(k)\,a^*(k)\,a(k) + \frac{g}{(2\pi)^{3/2}} \int dk\, \frac{[a(k) + a^*(-k)]}{\sqrt{2}\omega(k)^{1/2}}\,\tilde{j}(k). \tag{3.8}$$

[64] For a beautiful extensive discussion, see M. Reed and B. Simon, *Methods of Modern Mathematical Physics*, Vol. II (Fourier Analysis, Self-Adjointness), Academic Press 1975, Chap. X; for a sketchy account, see e.g. [SNS 96].

[65] A.S. Wightman, Introduction to some aspects of the relativistic dynamics of quantized fields, in *Cargèse Lectures in Theoretical Physics*, M. Levy ed., Gordon and Breach 1967, esp. Part II, Chap. VI; Constructive Field Theory. Introduction to the Problems, in *Fundamental Interactions in Physics and Astrophysics*, G. Iverson et al. eds., Plenum 1972. *Constructive Quantum Field Theory*, G. Velo and A.S. Wightman eds., Springer 1973.

[66] See J. Glimm and A. Jaffe, *Quantum Physics*, Springer 1981 and references therein.

[67] See e.g. [S 85] Part A, Sect. 2.3.

It is easy to see that the following "normal mode" operators

$$A(k) = a(k) + g\,(2\pi)^{-3/2}\,\tilde{\bar{j}}(k)\,\omega(k)^{-3/2}/\sqrt{2}$$

bring the Hamiltonian to the diagonal form

$$H = \int dk\,E(k)\,A^*(k)\,A(k) + E_0,$$

with $E(k) = (\mathbf{k}^2 + m^2)^{1/2}$ and

$$E_0 = -\tfrac{1}{2}g^2\,(2\pi)^{-3}\,(1/2)\int dk\,|\tilde{j}(k)|^2\,\omega(k)^{-2}.$$

If the current $j(k)$ does not decrease sufficiently fast when $k \to \infty$, as it happens for a point-like source (see below), E_0 is a divergent constant and it must be subtracted out by the addition of a suitable counterterm, in order to get a well defined Hamiltonian when the cutoffs are removed.

As we shall check below by an explicit calculation, the Fock representation for the normal mode operators A^*, A is also a Fock representation for a^*, a only if

$$\omega(k)^{-3/2}\,\tilde{j}(k) \in L^2(\mathbf{R}^3). \tag{3.9}$$

This condition may fail for UV reasons, namely if, for large $k, \tilde{j}(k) \to \text{const}$; this is what happens in the case of local interactions with a point-like source, $j(x) = \delta(x)$. The impossibility of having a Fock representation for both the time zero fields a^*, a and for the asymptotic fields A^*, A may also occur for IR reasons, namely if $\omega(k)^{-3/2}\,\tilde{j}(k)$ is not square integrable around $k = 0$. This is indeed what happens in the massless case, $m = 0$, if

$$Q \equiv \int dx\,j(x) = \tilde{j}(0) \neq 0.$$

This feature characterizes the Bloch-Nordsieck model of the infrared divergences of quantum electrodynamics (see below).

In both cases the ground state Ψ_0 of the total Hamiltonian, and the states of the representation defined by it, cannot be described in terms of the number of excitations, which are eigenstates of the free Hamiltonian, since

$$N = \int dk\,a^*(k)\,a(k)$$

does not exist as a well defined self-adjoint operator. In the case of mass gap, $m \neq 0$, also H_0 does not exist and in fact the Rayleigh-Schroedinger perturbative expansion is affected by divergences. For example, the expansion of Ψ_0 in terms of eigenstates of the free Hamiltonian would be

$$\Psi_0 = Z^{1/2}\sum_{n=0}^{\infty}\frac{1}{n!}\left(-\frac{g}{(2\pi)^{3/2}}\int dk\,\frac{\tilde{\bar{j}}(k)}{\sqrt{2}\,\omega^{3/2}}a^*(k)\right)^n\Psi_{0F}, \tag{3.10}$$

where Ψ_{0F} is the ground state of H_0, the Fock vacuum, and

$$Z = \exp\left[\frac{-g^2}{(2\pi)^3} \int dk \, \frac{|\tilde{j}(k)|^2}{2\omega(k)^3}\right]. \tag{3.11}$$

The integral in the exponent is divergent, and therefore Z vanishes if the condition of (3.9) does not hold.

It is worthwhile to remark that in this case for each value of the coupling constant g, one has an inequivalent representation, since the asymptotic fields $A_g, A_{g'}$, corresponding to two different values g, g' of the coupling constant are related by

$$A_{g'} = A_g + (g' - g) \, \tilde{j}(k)/[(2\pi)^{3/2} \sqrt{2}\,\omega(k)^{3/2}],$$

so that the Fock representation for A_g cannot also be so for $A_{g'}$, whenever (3.9) does not hold.

Example 3.2. The Bloch-Nordsieck model. The Bloch-Nordsieck (BN) model describes the (quantum) radiation field associated to a (classical) charged particle which moves with constant velocity $\mathbf{v}$ for $t < 0$ and with velocity $\mathbf{v}'$ for $t > 0$ (idealized scattering process). The equations of motion are

$$\Box \mathbf{A}(\mathbf{x}, t) = \mathbf{j}(\mathbf{x}, t), \tag{3.12}$$

which are equivalent to

$$i \, d\mathbf{a}(\mathbf{k}, t)/dt = \omega(k) \, \mathbf{a}(\mathbf{k}, t) + (2\omega)^{-1/2} \tilde{\mathbf{j}}(\mathbf{k}, t), \tag{3.13}$$

where

$$\mathbf{a}(\mathbf{k}, t) \equiv (2\omega)^{-1/2} \left[\omega(k)\mathbf{A}(\mathbf{k}, t) + i\dot{\mathbf{A}}(\mathbf{k}, t)\right], \quad \omega(k) = |\mathbf{k}|,$$

$$\mathbf{j}(\mathbf{x}, t) = e\,\mathbf{v}'\,\theta(t)\,\delta(\mathbf{x} - \mathbf{v}'t) + e\,\mathbf{v}\,\theta(-t)\delta(\mathbf{x} - \mathbf{v}t).$$

The solution is

$$\mathbf{a}(\mathbf{k}, t) = e^{-i\omega t} \left[e^{i\omega t_0} \mathbf{a}(\mathbf{k}, t_0) + (2\omega(k))^{-1/2} \int_{t_0}^{t} dt' \, e^{i\omega t'} \tilde{\mathbf{j}}(\mathbf{k}, t')\right]. \tag{3.14}$$

By taking the asymptotic limit $t_0 \to -\infty$ one gets the relation between the interacting field and the asymptotic in-field, e.g. for $t > 0$,

$$\mathbf{a}(\mathbf{k}, t) = e^{-i\omega t} \left(\mathbf{a}_{in}(\mathbf{k}) + \frac{e}{\sqrt{2\omega}} \left[\mathbf{v}' \frac{e^{i(\omega - \mathbf{k}\cdot\mathbf{v}')t} - 1}{\omega - \mathbf{k}\cdot\mathbf{v}'} + \mathbf{v}\frac{1}{\omega - \mathbf{k}\cdot\mathbf{v}}\right]\right)$$

$$\equiv e^{-i\omega t}\left[\mathbf{a}_{in}(\mathbf{k}) + \mathbf{f}(\mathbf{k}, t)\right].$$

Since $\mathbf{f}(\mathbf{k}, t) \notin L^2(\mathbf{R}^3)$ a Fock representation for $\mathbf{a}, \mathbf{a}^*$ cannot be a Fock representation for $\mathbf{a}_{in}, \mathbf{a}_{in}^*$ and conversely. In this (massless) case the existence of the free Hamiltonian for the asymptotic fields does not require a Fock representation for them.

The physical meaning of the above result is rather basic; in a scattering process of a charged particle the emitted radiation has a finite energy but an infinite number of "soft" photons, in the sense that for any finite ε the number of emitted photons with momentum greater than ε is finite, but the total number of emitted photons is infinite

$$\lim_{\varepsilon \to 0} \int_{|k| \geq \varepsilon} dk < \mathbf{a}^*(\mathbf{k}) \cdot \mathbf{a}(\mathbf{k}) >= \infty.$$

Such states with an infinite number of soft photons cannot be described in terms of an occupation number representation, but rather in terms of a classical radiation field $\mathbf{f}$ (which accounts for the low energy electromagnetic field) and hard photons. Such states are called coherent states and have been extensively studied in quantum optics.[68] The corresponding representation $\pi_{\mathbf{f}}$ of the creation and annihilation operators $\mathbf{a}^*, \mathbf{a}$ can be obtained from the Fock representation π_F by means of the following coherent transformation (*morphism*)

$$\rho(\mathbf{a}(\mathbf{k})) = \mathbf{a}(\mathbf{k}) + \mathbf{f}(\mathbf{k}), \quad \pi_{\mathbf{f}}(\mathbf{a}(\mathbf{k})) = \pi_F(\rho(\mathbf{a}(\mathbf{k}))),$$

where $\mathbf{f}$ is the classical radiation field.

The realization of the above basic (physical) mechanism, well displayed by the BN model, has led to the (non-perturbative) solution of the infrared problem in quantum electrodynamics. The charged (scattering) states define non-Fock coherent representations of the asymptotic electromagnetic algebra.[69] If this type of states are used to define the scattering amplitudes one gets finite results, when the infrared cutoff is removed, also in (the correspondingly adapted) perturbation theory.[70]

Other examples of non-Fock representations are provided by models with a non-vanishing ground state expectation value of the scalar field

$$\varphi(\mathbf{x}, 0) = \int dk\, e^{i\mathbf{k} \cdot \mathbf{x}} \, [\, a(\mathbf{k}) + a^*(-\mathbf{k}) \,](2\omega(k))^{-1/2},$$

since, if the representation is Fock for the canonical variables a, a^*, by Haag's argument the ground state must coincide with the Fock vacuum and the latter gives vanishing expectation of a and a^*.

---

[68] R.J. Glauber, Phys. Rev. Lett. **10**, 84 (1963); Phys. Rev. **131**, 2766 (1963); for an elementary account, see e.g. [S 85].

[69] V. Chung, Phys. Rev. **140B**, 1110 (1965); J. Fröhlich, G. Morchio and F. Strocchi, Ann. Phys. **119**, 241 (1979); G. Morchio and F. Strocchi, Nucl. Phys. **B211**, 471 (1984); for a review see G. Morchio and F. Strocchi, Erice Lectures, in *Fundamental Problems of Gauge Field Theory*, G. Velo and A.S. Wightman eds., Plenum 1986.

[70] T.W. Kibble, Phys. Rev. **173**, 1527; **174**, 1882; **175**, 1624 (1968) and references therein.

4 Mathematical Description
of Infinitely Extended Quantum Systems

From the discussion of the previous Chapter, it appears that the description of infinite systems looks much more difficult than in the finite dimensional case, above all because of the existence of (too) many possible representations of the algebra of canonical variables. A big step in the direction of controlling the problem has been taken by Haag et al., who emphasized the need of exploiting crucial physical properties of the algebra of observables in order to restrict their possible representations to the physically relevant ones. The crucial ingredient is the localization property of observable operations.

4.1 Local Structure

Any physically realizable operation is necessarily localized in space, since we cannot perform measurements or act on the system over the whole space. In order to encode this property in the structure of the algebra of observables, it is convenient to view it as generated by canonical variables or observables which have localization properties[71]. Thus, for each bounded space region V, one has the C^*-algebra $\mathcal{A}(V)$ of all observables (or canonical variables) localized in V.

A concrete realization of such a structure is obtained by considering canonical variables which have localization properties in the sense of (3.3). For regular test functions f, g of compact support contained in V, (typically $f, g \in \mathcal{D}(V)$), one considers the set of localized canonical variables

$$a(f) \equiv \int dx\ \psi(x)\ \bar{f}(x), \quad a^*(g) = \int dx\ \psi^*(x)\ g(x),$$

where $\psi(x)$ is a field "strictly localized in x" (see (3.3)). The algebra generated by such variables can be taken as the Heisenberg algebra localized in V. Similarly, a Weyl algebra localized in V is generated by the exponentials of

---

[71] For a general discussion of this strategy, see R. Haag, *Local Quantum Physics*, Springer 1996.

F. Strocchi: *Mathematical Description of Infinitely Extended Quantum Systems*, Lect. Notes Phys. **732**, 89–93 (2008)
DOI 10.1007/978-3-540-73593-9_4

the above localized canonical variables

$$U(f) = \exp\left[i\left(a(f) + a(f)^*\right)\right], \quad V(g) = \exp\left[a(g) - a(g)^*\right].$$

Quite generally, the association $V \to \mathcal{A}(V)$ realizes the identification of the algebras of observables localized in the volume V as V varies. The consistency of the physical interpretation requires that such a mapping satisfies the so called *isotony* property, namely $\mathcal{A}(V_1) \subseteq \mathcal{A}(V_2)$, whenever $V_1 \subseteq V_2$.

The physically motivated concept of localization has an algebraic translation in terms of commutation relations. For (equal time) space localization, the local structure of the algebras $\mathcal{A}(V)$ is formalized by the property

$$[\mathcal{A}(V), \mathcal{A}(V')] = 0, \quad \text{if} \quad V \cap V' = \emptyset. \tag{4.1}$$

For relativistic systems, it is more convenient to introduce algebras localized in bounded (open) space time regions $\mathcal{O}$ (usually taken as causally complete as it is the case of the diamonds or double cones[72]). Then, the locality property reads

$$[\mathcal{A}(\mathcal{O}_1), \mathcal{A}(\mathcal{O}_2)] = 0, \tag{4.2}$$

whenever $\mathcal{O}_2$ is spacelike with respect to $\mathcal{O}_1$, briefly $\mathcal{O}_2 \subset \mathcal{O}_1' \equiv$ the causal complement of $\mathcal{O}_1$. For observable algebras, this is the mathematical formulation of Einstein causality.[73]

The union of all $\mathcal{A}(V)$ (or $\mathcal{A}(\mathcal{O})$) is called the *local algebra*

$$\mathcal{A}_L \equiv \cup_\mathcal{V}\mathcal{A}(\mathcal{V}), \quad \mathcal{V} = V, \text{ or } = \mathcal{O}. \tag{4.3}$$

We have already argued before that it is convenient (if not necessary) to have a C^*-algebra and therefore one has to complete $\mathcal{A}_L$. As we shall see, this is a delicate point having deep connections with the dynamics and the physical description of the system. The most natural and simple choice is to consider the norm closure

$$\mathcal{A} \equiv \overline{\mathcal{A}_L}. \tag{4.4}$$

The norm closure leads to the smallest C^*-algebra generated by strictly local elements, all other topologies, like the (ultra-)strong and the (ultra-)weak being weaker, and therefore it gives the C^*-algebra with best localization properties. For this reason, the norm closure $\mathcal{A}$ is called the *quasi local algebra*.

Since the time evolution is one of the possible physically realizable operations, in order to have a consistent physical picture, the algebra of observables,

---

[72] A set $\mathcal{O}$ of points is *causally complete* if it coincides with its double causal complement, i.e. if $\mathcal{O} = (\mathcal{O}')'$, where $\mathcal{O}'$ (called the *causal complement* of $\mathcal{O}$) denotes the set of all points which are spacelike with respect to all points of $\mathcal{O}$.

[73] This concept of localization should not be confused with the problems discussed in connection with the Einstein-Podolski-Rosen paradox, see R. Haag, in *The Physicist's Conception of Nature*, J. Mehra ed., Reidel 1973; *Local Quantum Physics*, Springer 1992, p. 107; A.S. Wightman, in *Probabilistic Methods in Mathematical Physics*, F. Guerra et al. eds., World Scientific 1992.

and consequently its localization properties, must be stable under time evolution. We shall therefore take for granted that the time evolution defines a one-parameter group α_t, $t \in \mathbf{R}$ of *-automorphisms of the algebra of observables. Furthermore, we shall restrict our attention to systems for which also the space translations $\alpha_{\mathbf{x}}$, define *-automorphisms of the observable algebra.

For systems with a dynamics characterized by a finite propagation speed, the norm closure of $\mathcal{A}_L$ is stable under time evolution and therefore the quasi local algebra is a good candidate for the algebra of observables. This is the case of lattice spin systems with short range interactions[74] as well as the case of relativistic systems, since for them the causality requirement for the observables implies that under time evolution strictly local algebras are mapped into strictly local ones.

On the other hand, for non-relativistic systems the speed of propagation is in general infinite (even for the free Schroedinger propagator) and therefore some delocalization is unavoidable. Operators which are localized in a bounded region V at the initial time will not be so at any subsequent time. Therefore, the non-relativistic approximation necessarily requires a weaker form of locality, and, consequently, one should take as relevant algebra $\mathcal{A}$ a larger completion of $\mathcal{A}_L$[75]. We shall return to this point later.

4.2 Asymptotic Abelianess

Independently from the possible delocalization induced by the dynamics, strong physical reasons require that the algebra $\mathcal{A}$ of observables (or of the canonical variables) has at least the following (asymptotic) localization property, namely $\forall A, B \in \mathcal{A}$, putting $A_{\mathbf{x}} \equiv \alpha_{\mathbf{x}}(A)$,

$$\lim_{|\mathbf{x}| \to \infty} [A_{\mathbf{x}}, B] = 0. \tag{4.5}$$

Such a property is called *asymptotic abelianess* (in space). The physical meaning of such a property is rather transparent, since it states that the measurement of the observable A becomes compatible with the measurement of the observable B, in the limit in which A is translated at infinite space distance. Clearly, the validity of such a property for the algebra of observables is a necessary prerequisite for a reasonable quantum description of the corresponding system; otherwise, the measurement of the observable B would be influenced by possible measurements of observables at infinite space distances.

Asymptotic abelianess is obviously satisfied by local relativistic systems since in the limit $|\mathbf{x}| \to \infty$ the localization of $A_{\mathbf{x}}$ becomes space-like separated

---

[74] See O. Bratteli and D.W. Robinson, loc. cit. Vol. II, Sect. 6.2. For the convenience of the reader, a brief account is presented in the Appendix, Sect. 7.3.

[75] D.A. Dubin and G.L. Sewell, Jour, Math. Phys. **11**, 2290 (1970); G.L. Sewell, Comm. Math. Phys. **33**, 43 (1973); G. Morchio and F. Strocchi, Comm. Math. Phys. **99**, 153 (1985); J. Math. Phys. **28**, 622 (1987).

with respect to any fixed bounded region of space-time, and therefore the vanishing of the commutator is a consequence of Einstein causality.

Asymptotic abelianess is clearly satisfied by the local algebra $\mathcal{A}_L$ of a non-relativistic system, as a consequence of (4.1). It also holds for the quasi local algebra $\mathcal{A}$ defined as the norm closure of $\mathcal{A}_L$[76].

As stated in (4.5), asymptotic abelianess is an algebraic property (independent of the representation); from a physical point of view, it could be enough to require it to hold only in a class $\mathcal{F}$ of physically relevant representations and therefore the limit could be taken in the weak topology[77] defined by such representations

$$\text{w} - \lim_{|\mathbf{x}| \to \infty} [\pi(A_{\mathbf{x}}), \pi(B)] = 0, \quad \forall A, B \in \mathcal{A}, \ \forall \pi \in \mathcal{F}. \qquad (4.6)$$

In the sequel, we shall take for granted that the algebra $\mathcal{A}$ of observables (or of canonical variables) satisfies asymptotic abelianess, at least in the weak form of (4.6).

In a given representation π, the validity of the above equation extends to the case in which $\pi(B)$, $B \in \mathcal{A}$ is replaced by $B \in \overline{\pi(\mathcal{A})}^s$, where the bar with the suffix s denotes the strong closure[78].

In fact, $\forall \Psi, \Phi \in \mathcal{H}_\pi$ and $\forall \varepsilon > 0$, there exists a $B_1 \in \pi(\mathcal{A})$ such that $\|(B - B_1)\Phi\| \leq \varepsilon$, $\|(B - B_1)\Psi\| \leq \varepsilon$, so that

$$|(\Phi, [\pi(A_{\mathbf{x}}), B]\Psi)| \leq |(\Phi, [\pi(A_{\mathbf{x}}), B_1]\Psi)| + \varepsilon \|A\|(\|\Phi\| + \|\Psi\|)$$

---

[76] In fact, if $\mathcal{A}_L \ni A_n \to A$, $\mathcal{A}_L \ni B_n \to B$,

$$\| [A_{\mathbf{x}}, B] \| \leq \| [A_{n,\mathbf{x}}, B_m] \| + 2\|A_{n,\mathbf{x}}\| \|B - B_m\|$$

$$+ 2\|A_{\mathbf{x}} - A_{n,\mathbf{x}}\| (\|B_m\| + \|B - B_m\|)$$

and, since $\|A_{n,\mathbf{x}}\| = \|A_n\|$, in the limit $|\mathbf{x}| \to \infty$, the right hand side can be made as small as one likes.

[77] For the convenience of the reader, we recall that for the set $\mathcal{B}(\mathcal{H})$ of all bounded operators acting in the Hilbert space $\mathcal{H}$, the *weak topology* is defined by the seminorms given by the absolute values of the matrix elements of $\mathcal{B}(\mathcal{H})$ between vectors of $\mathcal{H}$, whereas the *strong topology* is given by the norms of the vectors $A\Psi$, $A \in \mathcal{B}(\mathcal{H})$, $\Psi \in \mathcal{H}$, (the *norm or uniform topology* is defined by the operator norm). Thus, for example, A_n converges weakly to A, (briefly $A_n \overset{w}{\to} A$), if, $\forall \Psi, \Phi \in \mathcal{H}$,

$$(\Psi, A_n \Phi) \to (\Psi, A\Phi).$$

On the other hand, A_n converges strongly to A, (briefly $A_n \overset{s}{\to} A$), if, $\forall \Psi \in \mathcal{H}$,

$$\|(A_n - A)\Psi\| \to 0.$$

[78] D. Kastler, in *Cargèse Lectures in Theoretical Physics*, Vol. IV, F. Lurçat ed., Gordon and Breach 1967, pp. 289-302.

and therefore in the limit $|\mathbf{x}| \to \infty$ the right hand side can be made as small as one likes.[79]

As we shall see below, the above property of asymptotic abelianess in the weak form (4.6) will play a crucial role in the analysis of the physically relevant representations of the observable algebra.

---

[79] By a similar argument one can also prove asymptotic abelianess when A and B are strong limits of elements of some $\mathcal{A}(V)$, on a common dense domain D stable under the implementers of the space translations.

5 Physically Relevant Representations

From the examples and the discussion of the previous chapter, it appears that for infinite systems the choice of the representation for the algebra of canonical variables (a basic preliminary step for even defining the dynamical problem) is a highly non-trivial problem (unless the model is exactly soluble). Among the possible representations of the relevant algebra $\mathcal{A}$, it is therefore convenient to isolate those which are physically acceptable. For the moment, we restrict our discussion to the zero temperature case. The non-zero temperature case will be briefly discussed in Chap. 12. On the basis of general physical considerations, we require the following conditions for a physically relevant representation π.

I. (**Existence of energy and momentum**) The space and time translations are described by strongly continuous groups of unitary operators $U(a)$, $U(t)$, $a \in \mathbf{R}^s$, $t \in \mathbf{R}$.
    By Stone's theorem, this guarantees the existence of the generators $\mathbf{P}$ (the momentum) and H (the energy), as well (densely) defined self-adjoint operators in the representation space $\mathcal{H}_\pi$. The existence of the energy is a necessary condition for the representation to be physically realizable. The implementability of the space translations is also necessary in relativistic quantum field theory, but could be dispensed with in many body theory and, e.g., be replaced by the invariance under a discrete subgroup of the translations. In the sequel, for simplicity, we shall not consider such more general cases.

II. (**Stability or spectral condition**) The spectrum $\sigma(H)$ of the Hamiltonian is bounded from below. The relativistically invariant form of the spectral condition is $\sigma(H) \geq 0$, $H^2 - \mathbf{P}^2 \geq 0$.
    Such a property guarantees that, under small (external) perturbations, the system does not collapse to lower and lower energy states.[80]

III. (**Ground state**) Inf $\sigma(H)$ is a (proper) non-degenerate eigen-value of the Hamiltonian. The corresponding eigenvector Ψ_0, called the *ground state*, has the following properties:

---

[80] This condition is not required for non-zero temperature states, since in that case the reservoir can feed the system and prevent it from collapsing.

F. Strocchi: *Physically Relevant Representations*, Lect. Notes Phys. **732**, 95–98 (2008)
DOI 10.1007/978-3-540-73593-9_5   © Springer-Verlag Berlin Heidelberg 2008

    i) Ψ_0 is a cyclic vector with respect to the local algebra

    ii) Ψ_0 is the unique translationally invariant state in $\mathcal{H}_\pi$.

Clearly, by a trivial redefinition of H, one can get $U(t)\Psi_0 = \Psi_0$.

The ground state condition is obviously satisfied in the free case, described by the Fock representation, with Ψ_0 being the Fock no-particle state. A physical justification for the existence of a ground state, in the general case, is that this is the state which the system should eventually reach, (when subject to small external perturbations), since the Hamiltonian is bounded from below.

From a mathematical point of view, the cyclicity of the ground state implies that the physically relevant representations can be obtained, through the GNS construction, from states (on the quasi local algebra) invariant under space and time translations, i.e. from correlation functions invariant under space and time translations.

From a physical point of view, the cyclicity requirement means that all the states of $\mathcal{H}_\pi$ can be approximated, as well as one likes, by local states, in agreement with the discussion in Sect. 4.1, i.e. the states of $\mathcal{H}_\pi$ can be described in terms of local operations on the ground state. In this picture, the ground state plays the role of the *reference state*, all the other states being essentially local modifications of it. This closely reflects the experimental limitation that, given a reference state, through physically realizable operations, one has access only to states which differ from it only locally.

Strictly speaking, the operational identification of the ground state involves some idealization or extrapolation, since one cannot actually measure or detect the properties of an infinitely extended system at space infinity. The identification of the ground state is therefore done on the basis of economy of the mathematical description, by extrapolating at infinity the large distance properties of the system. For example, in the case of a one-dimensional spin system, if all the relevant states (in a given phase) have the property that all the spins near the boundary point in the up direction, (as can be enforced by suitable boundary conditions), then, in the thermodynamical limit, the most economical description of such states of the system is in terms of (quasi) local modifications of an infinitely extended homogeneous state, in which all the spins are in the up direction.

In conclusion, the ground state completely accounts for the large distance behaviour of the system and this is the only ingredient which involves some extrapolation over the local character of the physically realizable operations.

The uniqueness of the translationally invariant state in any irreducible representation of $\mathcal{A}$ follows from asymptotic abelianess. The proof relies on Von Neumann's bicommutant theorem.[81] Given a *-subalgebra $\mathcal{A}$ of $\mathcal{B}(\mathcal{H})$ (the set of all bounded operators in $\mathcal{H}$), the *commutant*, denoted by $\mathcal{A}'$, is the set of all operators in $\mathcal{B}(\mathcal{H})$ which commute with $\mathcal{A}$, and the bicommutant (or double commutant) $\mathcal{A}'' \equiv (\mathcal{A}')'$ is the set of all operators in $\mathcal{B}(\mathcal{H})$ which

---

[81] For a sketch of the proof see e.g. [SNS 96].

commute with $\mathcal{A}'$. Clearly, if $\pi(\mathcal{A})$ is an irreducible representation of a C^*-algebra $\mathcal{A}$, then $\pi(\mathcal{A})' = \{\lambda \mathbf{1}, \lambda \in \mathbf{C}\}$ and $\pi(\mathcal{A})'' = \mathcal{B}(\mathcal{H})$.

Theorem 5.1. *(Von Neumann bicommutant). For a *-subalgebra $\mathcal{A}$ of $\mathcal{B}(\mathcal{H})$, with identity, the following three properties are equivalent*
i) $\mathcal{A} = \mathcal{A}''$
ii) $\mathcal{A}$ *is weakly closed (briefly* $\mathcal{A} = \overline{\mathcal{A}}^w$)
iii) $\mathcal{A}$ *is strongly closed:* $\mathcal{A} = \overline{\mathcal{A}}^s$.

Proposition 5.2. *In any irreducible representation π of the algebra $\mathcal{A}$ of observables, satisfying III i) and weak asymptotic abelianess, (4.6), the ground state is the unique translationally invariant state.*

Proof. In fact, if Ψ_0' is another translationally invariant state, which without loss of generality can be taken orthogonal to Ψ_0, $\forall A \in \mathcal{A}$, (denoting by P_0 the projection on Ψ_0), we have

$$(\Psi_0', \pi(A)\Psi_0) = (\Psi_0', \pi(A_\mathbf{x})\Psi_0) = (\Psi_0', \pi(A_\mathbf{x})P_0\Psi_0) =$$

$$(\Psi_0', P_0\,\pi(A_\mathbf{x})\Psi_0) + (\Psi_0', [\pi(A_\mathbf{x}), P_0]\Psi_0). \tag{5.1}$$

The first term on the right hand side is zero because Ψ_0' is orthogonal to Ψ_0. The second term is independent of $\mathbf{x}$ and one can take the limit $|\mathbf{x}| \to \infty$. Now, by Von Neumann's theorem and irreducibility

$$\overline{\pi(A)}^s = (\overline{\pi(A)}^s)'' \supseteq \pi(A)'' = \mathcal{B}(\mathcal{H}),$$

so that P_0 belongs to $\overline{\pi(A)}^s$ and therefore, by the extension of asymptotic abelianess (discussed after (4.6)), the second term vanishes in the limit $|\mathbf{x}| \to \infty$.

In conclusion, $(\Psi_0', \pi(A)\Psi_0) = 0$ and, by the cyclicity of Ψ_0, $\Psi_0' = 0$.

Under the same hypotheses, the above argument can be used to prove that

$$\mathrm{w} - \lim_{|\mathbf{x}| \to \infty} \pi(A_\mathbf{x}) = (\Psi_0, A\Psi_0)\mathbf{1} \equiv\, <A>_0 \mathbf{1}, \tag{5.2}$$

(sometimes, in the following equations the subscript 0 in the brackets will be omitted for simplicity). In fact, $\forall B \in \pi(\mathcal{A})$ one has, by asymptotic abelianess

$$\mathrm{w} - \lim_{|\mathbf{x}| \to \infty} \pi(A_\mathbf{x})\,B\,\Psi_0 = \mathrm{w} - \lim_{|\mathbf{x}| \to \infty} B\,\pi(A_\mathbf{x})\Psi_0 =$$

$$B\,\mathrm{w} - \lim_{|\mathbf{x}| \to \infty} ([\pi(A_\mathbf{x}), P_0]\Psi_0 + P_0\,\pi(A_\mathbf{x})\Psi_0).$$

By the extension of asymptotic abelianess, the first term on the right hand side vanishes and the second term is equal to $B\Psi_0 <A>_0$. Thus, the above weak limit exists and it equals the r.h.s. of (5.2).

The above equation (5.2) displays the fact that the ground state accounts for the large distance behaviour of the observables.

Since irreducibility of the GNS representation defined by a state Ω is equivalent to Ω being a pure state, the physically relevant irreducible representations of $\mathcal{A}$ are also called pure phases; in fact, they describe the pure phases of the system at zero temperature, in the standard sense of statistical mechanics. By the discussion of Sect. 4.1, they can also be interpreted as describing disjoint worlds.

Thus, as a consequence of asymptotic abelianess, which is crucially related to the local structure of the observables, different translationally invariant pure states on the observable algebra identify different phases or disjoint worlds, each characterized by different large distance (weak) limits of the observables.

6 Cluster Property and Pure Phases

The irreducible (physically relevant) representations selected in the previous section have a further important property, called *cluster property*.

Proposition 6.1. *Under the same conditions of Proposition 5.2, the ground state correlation of two quasi local operators factorize, when one is translated at space infinity*

$$\lim_{|\mathbf{x}| \to \infty} [< A B_{\mathbf{x}} >_0 - < A >_0 < B >_0] = 0 \qquad (6.1)$$

The proof follows easily from (5.2).

The reasons for stressing this property are many. First, the cluster property plays a crucial role for the foundations of the S-matrix theory in quantum field theory.[82] In fact, the possibility itself of defining a scattering process requires such a factorization of the amplitude relative to clusters of fields which are infinitely separated in space. Otherwise, a scattering process localized in a space time region $\mathcal{O}$ would be influenced by a scattering taking place at very large distances.

The physical meaning of the cluster property is that the ground state reacts locally to local operations, and it cannot support non-trivial correlations between far separated observables. In a certain sense, this condition neutralizes the non-local content of the ground state to the effect that the latter does not spoil the local structure of the physically realizable operations, at the level of the correlation functions, and it is essentially confined to the property of accounting for the large distance limits of the observables.

For representations satisfying conditions I, II, III i), one can show that the cluster property implies irreducibility and therefore it is equivalent to it, but, from a constructive point of view, the cluster property is much better controlled since it can be directly read off from the knowledge of the correlation functions. Thus, for zero temperature states the cluster property can be used to identify the pure phases. Actually, as we shall see later, the cluster

---

[82] R. Haag, Phys. Rev. **112**, 669 (1958); *Local Quantum Physics*, Springer 1996, esp. Sect. II.4; D. Ruelle, Helv. Phys. Acta **35**, 147 (1962). For a systematic account of the Haag-Ruelle theory, see R. Jost, *The General Theory of Quantized Fields*, Am. Math. Soc. 1965.

F. Strocchi: *Cluster Property and Pure Phases*, Lect. Notes Phys. **732**, 99–103 (2008)
DOI 10.1007/978-3-540-73593-9_6 © Springer-Verlag Berlin Heidelberg 2008

property characterizes the pure phases also at non-zero temperature, where irreducibility cannot hold.

The important property which makes the cluster property so relevant (also at non-zero temperature) is that of being equivalent to the uniqueness of the translationally invariant state. In particular, this condition (assumed in III) appears justified also on the basis of the motivations for the validity of the cluster property discussed above and, in fact, can be replaced by the latter.

For the proof of such an equivalence, we remark that the cluster property in the form of (6.1) states both the existence of the limit of the first term and the property of being equal to the second. In order to make such a relation more transparent and to point out its basic physical content, we shall first discuss a weaker form of the cluster property in which the limit in (6.1) is taken in the Cesaro sense (*weak cluster property*), and we shall prove the equivalence between such a weak form and the uniqueness of the translationally invariant state.

We recall that such a weaker form of the limit (for brevity also called *mean-limit* or *mean ergodic limit*) is defined in the following way, for locally measurable functions (for simplicity we consider the case of one variable):

$$\text{mean} - \lim_{|x| \to \infty} f(x) \equiv \lim_{L \to \infty} L^{-1} \int_0^L dx \, f(x). \tag{6.2}$$

The limit can easily be proved to exist for a large class of functions, e.g. if the Fourier transform of f is a finite measure. It is clear that the values taken by f in any bounded interval $[0, L_0]$ do not affect the right hand side since the latter is also equal to

$$\lim_{L \to \infty} L^{-1} \int_{L_0}^L dx \, f(x).$$

The only thing which matters for the limit is the behaviour of f at infinity and clearly, if $f(x)$ has a limit in the ordinary sense, the mean-limit coincides with it.

Theorem 6.2. *In any representation π defined by a translationally invariant state and satisfying weak asymptotic abelianess, the weak cluster property,*

$$\lim_{|V| \to \infty} |V|^{-1} \int_V dx \, [< A \, B_{\mathbf{x}} > - < A >< B >] = 0, \tag{6.3}$$

where V is a bounded (regular) region centered at the origin, e.g. a sphere or a cube, $|V|$ denotes the volume of V and the limit is taken by expanding it equally in all directions, is equivalent to the uniqueness of the translationally invariant state.

Proof. The proof exploits the continuous version of Von Neumann's ergodic theorem,[83] according to which, if $U(x)$ is a group of unitary (translation) operators in a Hilbert space $\mathcal{H}$,

$$\text{mean} - \lim_{x \to \infty} U(x) = P_{inv}, \qquad (6.4)$$

where P_{inv} denotes the projection on the subspace $\mathcal{H}_{inv}$ of $U(x)$ invariant vectors (and the limit exists in the strong topology)[84]. A simple consequence[85] of such a theorem is that

$$\lim_{|V| \to \infty} |V|^{-1} \int_V dx\, U(x) = P_{inv}, \qquad (6.5)$$

with P_{inv} the projection on the subspace of vectors, which are invariant under $U(x)$, $\forall x \in \mathbf{R}^s$. It then follows trivially that (6.3) holds iff P_{inv} is one-dimensional, i.e. there is only one state invariant under space translations and therefore $P_{inv} = P_0$ (the projection on the ground state).

In order to get the equivalence between the uniqueness of the translationally invariant state and the cluster property in the (strong) form of (6.1), one has to control the limit $|\mathbf{x}| \to \infty$. The existence of such a limit in the weak sense is guaranteed if the representation has the property that the *center*

$$\mathcal{Z} \equiv \pi(\mathcal{A})' \cap \pi(\mathcal{A})''$$

is pointwise invariant under space translations.

The pointwise invariance of the center under space translations follows from the relativistic spectral condition[86], and one may argue about its validity

---

[83] See e.g. M. Reed and B. Simon, *Methods of Modern Mathematical Physics*, Vol. I, Academic Press 1972, Sect. 11.5.

[84] The point is that oscillatory behaviours are killed by the mean limit and only the zero frequency part survives. We briefly sketch the proof. Equation (6.4) trivially holds on $\mathcal{H}_{inv}$, so that by the linearity of $U(x)$ it remains to check it on $\mathcal{H}_{inv}^{\perp}$, which is equal to the closure of $\{(\mathbf{1} - U(x))\mathcal{H}, x \in \mathbf{R}\}$, since $U(x)\Psi = \Psi$ implies $U(x)^*\Psi = \Psi$ and for a vector Ψ, the condition of being in $\mathcal{H}_{inv}$, i.e. $((\mathbf{1} - U(x)^*)\Psi, \mathcal{H}) = 0, \forall x$, is equivalent to $(\Psi, (\mathbf{1} - U(x))\,\mathcal{H}) = 0, \forall x$. Now, for vectors of the form $\Psi = (\mathbf{1} - U(y))\,\Phi$, the integral occurring in the mean limit reads

$$\int_0^L dx\,(U(x) - U(x+y))\,\Phi = \left(\int_0^L - \int_y^{L+y}\right)dx\,U(x)\,\Phi = \left(\int_{V_1} - \int_{V_2}\right)dx\,U(x)\,\Phi,$$

where $V_1 \equiv [0, L] \setminus ([y, L+y] \cap [0, L])$, $V_2 \equiv [y, L+y] \setminus ([y, L+y] \cap [0, L])$. Then, the norm of the l.h.s. of (6.4) applied to $\Psi = (\mathbf{1} - U(y))\,\Phi$ is bounded by

$$L^{-1} |([0, L] \cup [y, L+y])/([y, L+y] \cap [0, L])| \, ||\Phi|| \xrightarrow[L \to \infty]{} 0.$$

[85] It suffices to apply the theorem to each variable $x_i, i = 1, 2, ..., s$, by e.g. integrating $U(x_1, x_2, ..., x_s)$ over $V_1 \times V_2 \times ...V_s$.

[86] H. Araki, Prog. Theor. Phys. **32**, 884 (1964).

for non-relativistic systems. The important physical property following from it is that the existence of the weak limits imply that they coincide with the ergodic averages

$$w - \lim_{V \to \infty} V^{-1} \int_V dx \, \alpha_{\mathbf{x}}(A), \tag{6.6}$$

which describe macroscopic observables (for simplicity V denotes both the bounded region and its volume).

A special case in which the pointwise (space translation) invariance of the center obviously holds is that of the so-called *factorial representations*, defined by the condition of having a trivial center: $\mathcal{Z} = \{\lambda \mathbf{1}, \lambda \in \mathbf{C}\}$. The class of factorial representations includes in particular the irreducible representations (for which $\pi(\mathcal{A})' = \{\lambda \mathbf{1}, \lambda \in \mathbf{C}\}$), but is much more general; in fact, it can be taken as the mathematical characterization of the pure phases, also at non-zero temperature (where the representation cannot be irreducible). The physical motivation for such a choice is that the ergodic decomposition of a representation with respect to the space translations[87] automatically leads to definite values for the macroscopic observables.

Proposition 6.3. *In any representation defined by a translationally invariant state, satisfying weak asymptotic abelianess, with the property that the center $\mathcal{Z}$ is pointwise invariant under translations, one has*

$$w - \lim_{|\mathbf{x}| \to \infty} A_{\mathbf{x}} B \Psi_0 = B \, P_{inv} \, A \Psi_0, \tag{6.7}$$

where P_{inv} denotes the projection on the subspace of translationally invariant vectors and the symbols A, B denote the representatives of elements of $\mathcal{A}$ in the given representation.

Proof. One can essentially use the same argument as in the derivation of (5.2), with P_0 replaced by the projection P_{inv}. In fact, $\forall B \in \pi(\mathcal{A})$,

$$A_{\mathbf{x}} B \Psi_0 = \{[A_{\mathbf{x}}, B] + B [A_{\mathbf{x}}, P_{inv}] + B P_{inv} A_{\mathbf{x}}\} \Psi_0 \tag{6.8}$$

and in the limit $|\mathbf{x}| \to \infty$, the first term vanishes by asymptotic abelianess, and the last term is independent of $\mathbf{x}$.

Thus, one has to discuss the weak limit of the second term. To this purpose, one notes that $\|A_{\mathbf{x}}\| = \|A\|$, since the space translations are automorphisms of $\mathcal{A}$ and therefore norm preserving. Then, by a compactness argument there are subsequences $\{A_{\mathbf{x}_n}\}, |\mathbf{x}_n| \to \infty$, which have weak limits $z_{\{\mathbf{x}_n\}}$. By asymptotic abelianess such weak limits belong to the center $\mathcal{Z}$ and (by hypothesis) commute with the spectral projections of $U(\mathbf{x})$, in particular with P_{inv}. Thus, for all convergent subsequences

$$w - \lim_{|\mathbf{x}_n| \to \infty} [A_{\mathbf{x}_n}, P_{inv}] = 0. \tag{6.9}$$

---

[87] See O. Bratteli and D.W. Robinson, *Operator Algebras and Quantum Statistical Mechanics*, Vol. 1. Springer 1987, Sect. 4.3.

This implies that the second term in (6.8) converges to zero, since otherwise there is an $\varepsilon > 0$, a pair Ψ, $\Phi \in \mathcal{H}$ and a sequence $\{\mathbf{y}_n\}$, $|\mathbf{y}_n| \to \infty$, such that $(\Psi, [A_{\mathbf{y}_n}, \mathcal{P}_{inv}] \Phi)| > \varepsilon$ for all $|\mathbf{y}_n|$ sufficiently large; by the compactness argument the sequence $\{A_{\mathbf{y}_n}\}$ has a convergent subsequence which would therefore not satisfy (6.9). In conclusion, the weak limit $|\mathbf{x}| \to \infty$ of the left hand side of (6.8) exists and (6.7) holds.

Proposition 6.4. *In a representation defined by a translationally invariant state, satisfying weak asymptotic abelianess, with the center pointwise invariant under space translations, the cluster property (6.1) is equivalent to the uniqueness of the translationally invariant state.*

In a factorial representation the translationally invariant state is unique.

Proof. It follows easily from (6.7) that the cluster property holds iff P_{inv} is one-dimensional. For factorial representation (6.7) holds with P_{inv} replaced by the projection P_0 on the translationally invariant vector state Ψ_0, which defines the representation, since the (trivial) center obviously commutes with P_0 and therefore the cluster property holds.

It is worthwhile to remark that irreducibility is a much too strong condition for non-isolated systems, like those in thermodynamical equilibrium at non-zero temperature, which require a heat exchange with the reservoir (or thermal bath). The GNS representation defined by a translationally invariant equilibrium state has the property that the equilibrium vector state is cyclic with respect to the observable algebra, but there are operators (e.g. those describing the "dynamical variables" of the reservoir), which commute with the observables of the system and therefore irreducibility fails. However, if the representation is factorial, by Proposition 6.4 the translationally invariant equilibrium state cannot be decomposed as a convex combination of other traslationally invariant states and in this sense describes a pure phase. We shall return to non-zero temperature states later.

The physically motivated factorization of the correlation functions of infinitely separated observables does not require irreducibility, but rather the uniqueness of the translationally invariant state, which holds if the representation is factorial.

As it is clear from the above discussion, in a factorial representation defined by a traslationally invariant equilibrium state, the ergodic averages of observables are c-numbers and coincide with the expectation values of the observables on the equilibrium state. Such a state therefore encodes the information on the macroscopic observables, as well as the large distance behaviour of the observables.

In the next section we shall confront the general framework discussed above with some concrete examples.

7 Examples

7.1 Spin Systems with Short Range Interactions

As mentioned before, the quantum mechanics of infinite systems is not under mathematical control as it is in the finite dimensional case. A non-perturbative control has been achieved for quantum field theories in low space-time dimensions ($d = 1+1$, $d = 2+1$), but the question is still open in $d = 3+1$ dimensions and the triviality of the φ^4 theory indicates that the perturbative expansion is not reliable for existence problems. It is clear that the existence of a non-trivial dynamics for systems with infinite degrees of freedom is not a trivial problem, but for non-relativistic systems some result is available. As a matter of fact, for spin systems with short range interactions the infinite volume dynamics α_t has been shown to exist.[88]

To give an idea of how the problem is attacked and solved, we first consider the simple case of a one-dimensional chain of spins (a one-dimensional "ferromagnet") with a formal Hamiltonian of Ising type

$$H = -J \sum_{i,j} \sigma_i \sigma_j, \quad J > 0, \tag{7.1}$$

where the sum is over all the nearest neighbor pairs of indices i, j, which denote the chain sites and σ denotes the component of the spin along the z direction.

The local algebra of observables is generated by the spin operators σ at the various sites; in particular, for each volume V the algebra $\mathcal{A}(V)$ is the algebra generated by the spins sitting in the sites $i \in V$.[89] Since spins at different sites are assumed to commute, the localization condition (4.1) obviously holds and asymptotic abelianess is satisfied by the quasi local algebra $\mathcal{A}$ (the norm closure of the local algebra).

The first non-kinematical question is the existence of the time evolution and the stability of $\mathcal{A}$ under it. To this purpose, as discussed before, we replace the formal (ill defined) Hamiltonian (7.1) with the (infrared regularized) finite

---

[88] D.W. Robinson, Comm. Math. Phys. **7**, 337 (1968).

[89] For a detailed discussion of the mathematical structure of spin models, see O. Bratteli and D.W. Robinson, loc. cit. Vol. 2, Sect. 6.2.

F. Strocchi: *Examples*, Lect. Notes Phys. **732**, 105–113 (2008)
DOI 10.1007/978-3-540-73593-9_7 © Springer-Verlag Berlin Heidelberg 2008

volume Hamiltonian

$$H_V \equiv -J \sum_{(i,j) \in V} \sigma_i \, \sigma_j, \tag{7.2}$$

which is well defined since it involves only a finite number of terms. Then, we consider the finite volume dynamics

$$\alpha_t^V (A) \equiv e^{iH_V t} \, A \, e^{-iH_V t}, \quad A \in \mathcal{A}_L \tag{7.3}$$

and try to define the infinite volume dynamics as a limit of α_t^V when $V \to \infty$.

The idea is that for $A \in \mathcal{A}(V_0)$, V_0 fixed, the above transformation $\alpha_t^V (A)$ becomes independent of V for V large enough, thanks to (4.1) and the nearest neighbor coupling. Thus, the limit of $\alpha_t^V (A)$ exists in norm and it defines a time evolution α_t as an automorphism of the local algebra, which can be extended to an automorphism of the quasilocal algebra $\mathcal{A}$, since it is norm preserving. The existence of the time evolution as a norm limit of finite volume dynamics has been proved quite generally for any lattice spin Hamiltonian with short range interactions, i.e. with absolutely summable spin interaction potentials.[90] For example, in $d = 3$ space dimensions, such short range interactions include Hamiltonians of the form

$$H = \sum_{i,j} J_{ij} \, \boldsymbol{\sigma}_i \cdot \boldsymbol{\sigma}_j.$$

provided the potential J_{ij} decreases at least as $|i - j|^{-3-\varepsilon}$, $\varepsilon > 0$.

Essentially the same logic applies to Hamiltonians which are local functions of canonical variables or fields, which satisfy the locality condition (4.1) or (4.2), respectively. This is the case of UV regularized quantum field theories, for which the infinite volume limit of the time evolution of local operators can be proved to exist by locality.[91]

7.2 Free Bose Gas. Bose-Einstein Condensation

The Bose-Einstein condensation is the very important collective effect at the basis of the phenomenon of superfluidity[92] and it provides an interesting example of an infinite system also at the level of the *free Bose gas*.

---

[90] D.W. Robinson, Comm. Math. Phys. **7**, 337 (1968). For the convenience of the reader, also due to the conceptual relevance of the result, a sketch of the proof is given in Appendix 7.3.

[91] M. Guenin, Comm. Math. Phys. **1**, 127 (1966); I.E. Segal, Proc. Natl. Acad. Sci. USA, **57**, 1178 (1967).

[92] For a simple account see [S 85].

The model is defined[93] by the Weyl algebra $\mathcal{A}$ generated by the essentially localized field operators $\psi(f)$, $\psi(g)^*$, $f, g \in \mathcal{S}(\mathbf{R}^s)$ (see the discussion in Sect. 4.1),

$$\psi(f) = \int d^s x\, \psi(x)\, f(x),$$

with

$$[\psi(x), \psi(y)^*] = \delta(x - y), \quad [\psi(x),\, \psi(y)] = 0.$$

The formal Hamiltonian describing a system of free bosons is

$$H = (1/2m) \int d^s x\, |\boldsymbol{\nabla}\, \psi(x)|^2.$$

It is (formally) positive, so that if a state is annihilated by H, (more precisely by any finite volume restriction H_V of H), it is a lowest energy state. The condition $H_V \Psi_0 = 0$, $\forall V$ implies

$$\boldsymbol{\nabla}\psi(x)\Psi_0 = 0, \ \forall x, \tag{7.4}$$

which must be solved compatibly with the condition that one has finite density.[94] Equation (7.4) can be written as

$$0 = -i\boldsymbol{\nabla}\psi(x)\Psi_0 = [\mathbf{P},\, \psi(x)]\Psi_0 = \mathbf{P}\,\psi(x)\,\Psi_0,$$

where we have required the translational invariance of Ψ_0. The uniqueness of the translationally invariant state requires

$$\psi(x)\,\Psi_0 = c\,\Psi_0, \ c = (\Psi_0,\, \psi(x)\,\Psi_0) \equiv\, < \psi >, \tag{7.5}$$

(a smearing with test functions would give a mathematically precise meaning to the above equations).[95] Therefore, the ground state defines a Fock representation for the operators ψ_F, ψ_F^* defined by

$$\psi_F(x) \equiv \psi(x) - < \psi >$$

---

[93] For a rigorous mathematical treatment see H. Araki and E.J. Woods, J. Math. Phys. **4**, 637 (1963); D.A. Dubin, *Solvable models in algebraic statistical mechanics*, Claredon Press, Oxford 1974. See also N.M. Hugenholtz, in *Fundamental Problems in Statistical Mechanics II*, E.G.D. Cohen ed., North-Holland, Amsterdam 1968, p. 197 and O. Bratteli and D.W. Robinson, loc. cit. Vol. 2, Sect. 5.2.5.

[94] To make the argument mathematically rigorous, one can solve the problem in a finite volume with periodic boundary conditions and then take the thermodynamical limit, as discussed in the references of the previous footnote.

[95] Equation (7.5) is incompatible with canonical anti-commutation relations, and in fact, as it is well known, the ground state for a free Fermi gas is not annihilated by the above free Hamiltonian.

and one can easily compute the correlation functions of ψ, ψ^* in terms of those of $\psi_F, \psi_F^*,$[96] e.g.

$$< \psi(x)^* \, \psi(y) > = | < \psi > |^2, \quad < \psi(x) \, \psi(y) > = < \psi >^2, \quad \text{etc.}$$

From the above equations it follows that $< \psi >$ is related to the average density

$$| < \psi > |^2 = < \psi(x)^* \, \psi(x) > = n, \quad < \psi > = \sqrt{n} \, e^{i\theta}, \ \theta \in [0, 2\pi).$$

The ground state can be thought as labeled by the "order parameter" $< \psi >$ and, in order to spell this out, we shall denote the ground state by $\Psi_{0,n,\theta}$ or, briefly, by Ψ_θ and the corresponding state on $\mathcal{A}$ by Ω_θ, (θ *ground state*).

For any θ the GNS representation defined by Ω_θ is irreducible because the algebras generated by ψ, ψ^* and by ψ_F, ψ_F^* coincide and the latter is irreducible. It is not difficult to see that different values of $< \psi >$ label inequivalent representations of $\mathcal{A}$ (see also the discussion below). Properties I-III of Chap. 5 are obviously satisfied.

The *localization* properties of the model deserve a few comments, since we have a very simple example of the conceptual problem of identifying an algebra with localization properties stable under time evolution. As a matter of fact, the quasi local algebra obtained as the norm closure of the local algebra $\mathcal{A}_L$, generated by the Weyl exponentials $U(f)$, $V(g)$, $f, g \in \mathcal{D}(\mathbf{R}^s)$ (see Sect. 4.1), is not stable under time evolution. The point is that the Schroedinger time evolution does not map $\mathcal{D}(\mathbf{R}^s)$ into $\mathcal{D}(\mathbf{R}^s)$ and, therefore, strictly localized operators at $t = 0$ are no longer so at any subsequent time. This implies that

$$\alpha_t(U(f)) = U(f_t), \quad \tilde{f}_t(k) \equiv \tilde{f}(k) \, e^{ik^2 \, t/2m},$$

is not in the norm closure of $\mathcal{A}_L$.[97] Thus, the non-relativistic approximation and the corresponding time evolution require to weaken slightly the condition of localization by replacing the local algebra $\mathcal{A}_L$, e.g. by the essentially local

---

[96] They provide much more detailed information than the mere probability distribution of the occupation numbers, as it is done in the standard elementary treatments of the free Bose gas: see e.g. R.P. Feynman, *Introduction to Statistical Mechanics*, Benjamin 1972, Sect. 1.9.

[97] D.A. Dubin and G.L. Sewell, J. Math. Phys. **11**, 2990 (1970); G.L. Sewell, Comm. Math. Phys. **33**, 43 (1973). The point is that

$$\|U(f_t) - U(g_n)\| = \|e^{i \operatorname{Im}(f_t, \, g_n)} \, U(f_t - g_n) - \mathbf{1}\| = 2,$$

unless $\|f_t - g_n\|_{L^2} = 0$, because $A \equiv \psi(h) + \psi(h)^*$ is an unbounded operator with continuous spectrum (linear in h) and therefore

$$\|e^{i \, A} - \mathbf{1}\|^2 = \sup_{\lambda \in \sigma(A)} |e^{i\lambda} - 1|^2 = 4.$$

On the other hand, if $f, g_n \in \mathcal{D}(\mathbf{R}^s)$, $\|f_t - g_n\|_{L^2}$ cannot vanish, since $f_t \notin \mathcal{D}$.

algebra $\mathcal{A}_l$, generated by the Weyl exponentials of $\psi(f)$, $\psi(g)^*$, $f, g \in \mathcal{S}(\mathbf{R}^s)$, which is stable under time evolution. It is easy to see that $\mathcal{A}_l$ and its norm closure $\mathcal{A}$ satisfy asymptotic abelianess.

Another interesting feature of the model is related to *gauge invariance*. The *gauge transformations*

$$\beta^\lambda(\psi(x)) = e^{i\lambda}\psi(x), \quad \beta^\lambda(\psi^*(x)) = e^{-i\lambda}\psi^*(x), \quad \lambda \in [0, 2\pi]$$

define a one-parameter group of *-automorphisms of $\mathcal{A}$, which commutes with α_t. The ground state $\Psi_{0,\theta}$ is not gauge invariant, in the sense that its correlation functions are not invariant under β^λ, since e.g.

$$< \beta^\lambda(\psi) >_\theta = < \psi >_{\theta+\lambda} .$$

In fact, under gauge transformations $\Omega_\theta \to \Omega_{\theta+\lambda}$.

A gauge invariant state can be defined by averaging over θ

$$\Omega(A) \equiv (2\pi)^{-1} \int_0^{2\pi} d\theta \, \Omega_\theta(A), \quad \forall A \in \mathcal{A}. \tag{7.6}$$

One has

$$\Omega(\psi_1^*...\psi_k^*\psi_{k+1}...\psi_{k+j}) = (2\pi)^{-1} \int_0^{2\pi} d\theta \, (n)^{(k+j)/2} \, e^{-i(k-j)\theta},$$

which vanishes unless $k = j$; similarly for $A = \psi_1...\psi_k\psi_{k+1}^*...\psi_{k+j}^*$ one has $\Omega(A) = 0$, if $k \neq j$, since $\Omega_\theta(A) = \Omega_0(\beta^\theta(A)) = e^{i\theta(k-j)} \Omega_0(A)$.

As displayed by (7.6), Ω is not a pure state on $\mathcal{A}$ and the GNS representation defined by Ω is not irreducible. This can be explicitly seen by noting that

$$\psi(f)_\infty = \lim_{V\to\infty} V^{-1} \int_V d^s x \, (\psi(f))_x$$

commutes with $\mathcal{A}$, by asymptotic abelianess, and $\exp i(\psi(f))_\infty$ belongs to the centre $\mathcal{Z}$; on the other hand

$$\Omega((\psi(f))_\infty) = 0, \quad \Omega((\psi(f)^*)_\infty \, (\psi(f))_\infty) = n,$$

so that $\exp i(\psi(f))_\infty$ is not a multiple of the identity in the GNS representation defined by Ω and this excludes irreducibility.

A simple computation gives

$$\Omega_\theta((\psi(f))_\infty) = \sqrt{n} e^{i\theta} \tilde{f}(0), \quad \Omega_\theta(e^{i(\psi(f)+\psi(f)^*)_\infty}) = e^{2iRe\,(<\psi>_\theta \tilde{f}(0))},$$

so that for different θ the states Ω_θ assign different values to an element of the centre and therefore the corresponding representations are inequivalent.

The algebra $\mathcal{A}$ contains a (pointwise) gauge invariant subalgebra $\mathcal{A}_{obs}$, which has the meaning of the algebra of observables. All the states Ω_θ and therefore Ω define equivalent representations of $\mathcal{A}_{obs}$.

The ground state correlation functions for the *free Fermi gas* can be computed by putting the system in a box of volume V with periodic boundary conditions. The ground state Ω_V is completely characterized by the two point function

$$\Omega_V(a_k^* \, a_{k'}) = \delta_{k,k'} \, \theta(k_F^2 - k^2), \quad k_F^3 \equiv 3\pi^2 \, n, \tag{7.7}$$

where θ denotes the Heaviside step function and n the density. Thus, in the thermodynamical limit

$$< \psi(f)^* \, \psi(g) >_{\Omega_V} = V^{-1} \sum_j \overline{\tilde{f}}(k_j) \, \tilde{g}(k_j) \, \theta(k_F^2 - k_j^2)$$

$$\rightarrow (2\pi)^{-3} \int d^3k \, \overline{\tilde{f}}(k) \, \tilde{g}(k) \, \theta(k_F^2 - k^2). \tag{7.8}$$

7.3 * Appendix: The Infinite Volume Dynamics for Short Range Spin Interactions

We consider a spin system on a lattice Z^d with many-body "potentials"

$$\Phi^k(x_1, ...x_k) = v(x_1, ...x_k)\sigma(x_1)...\sigma(x_k),$$

where x_i denote the lattice points and for simplicity the spin components are not spelled out. Briefly, if $X = \{x_1, ...x_k\}$ denotes a set of lattice points, we denote by $\Phi(X)$ the corresponding interaction energy. For example, in the case of a spin system interacting only via a two body potential, one has $\Phi(X) = 0$, unless $X = \{x_1, x_2\}$ and $\Phi(X) = J(x_1, x_2)\sigma(x_1) \, \sigma(x_2)$.

The potentials are assumed to describe translationally invariant interactions, i.e.

$$\alpha_a(\Phi(X)) = \Phi(X + a).$$

The interaction is said to be of *finite range* if, given a lattice point x, the number of sets X, which contain x and for which $\Phi(X) \neq 0$, is finite; the union of such sets is denoted by Δ and called the range of Φ; $N(\Delta)$ denotes the number of points of Δ.

The finite volume Hamiltonian is therefore of the following form

$$H_V = \sum_{X \subset V} \Phi(X).$$

We shall first consider the case of finite range and sketch the proof[98] that, $\forall A \in \mathcal{A}_L$, $\alpha_t^V(A)$ converges in norm. This implies that the norm limit α_t is norm preserving: $||\alpha_t(A)|| = ||A||$ and therefore it defines an automorphism of $\mathcal{A}_L$. Thus, α_t can be extended to the norm closure $\mathcal{A}$ of $\mathcal{A}_L$, the extension

---

[98] D.W. Robinson, Comm. Math. Phys. **7**, 337 (1968); R.F. Streater, Comm. Math. Phys. **6**, 233 (1967).

is norm preserving and it leaves $\mathcal{A}$ stable.[99] Hence, the dynamics exists as an automorphism of $\mathcal{A}$.

In order to prove the norm convergence, $\forall\, A \in \mathcal{A}_L$ we consider

$$\alpha_t^V(A) = e^{itH_V}\, A\, e^{-itH_V} = A + it[H_V, A] + ... = \sum_{n=0}^{\infty} A_n^V t^n, \qquad (7.9)$$

$$A_n^V \equiv (i^n/n!) \sum_{X_1,...X_n \subset V} [\Phi(X_n), [\Phi(X_{n-1}), ...[\Phi(X_1), A]...]]. \qquad (7.10)$$

The finite range implies that, for fixed n, the r.h.s. of (7.10) becomes independent of V, for V large enough, i.e. the series (7.9) is convergent term by term and we only need an estimate on $A_n = \lim A_n^V$ to get the convergence of the series. For this purpose, we consider the multiple commutator $B_n(A)$ appearing on the r.h.s. of (7.10). If $A \in \mathcal{A}(V_0)$, one has that $B_n(A) \in \mathcal{A}(V_1)$,

$$V_1 \equiv X_{n-1} \cup X_{n-2} \cup ... \cup V_0, \quad N(V_1) = N(V_0) + (n-1)N(\Delta).$$

Hence, by locality $[\Phi(X), B_n] = 0$, if $X \cap V_1 = \emptyset$, and, by using translation invariance, we get

$$\|\sum_{X_n \subset V}[\Phi(X_n), B_n]\| \leq \sum_{X_n \subset V,\, X_n \cap V_1 \neq \emptyset} \|[\Phi(X_n), B_n]\|$$

$$\leq 2\|B_n\| \sum_{X \subset V,\, X \cap V_1 \neq \emptyset} \|\Phi(X)\| = 2\|B_n\|\, N(V_1) \sum_{X \ni 0} \|\Phi(X)\|.$$

Then, by iteration, we get

$$\|A_n^V\| \leq n!^{-1}\|A\| \left(2\sum_{X \ni 0} \|\Phi(X)\|\right)^n \prod_{k=0}^{n-1}(N(V_0) + kN(\Delta))$$

$$\leq n!^{-1}\left(2\sum_{X \ni 0}\|\Phi(X)\|\right)^n (N(V_0) + nN(\Delta))^n\|A\|$$

$$\leq \|A\|\left(2\sum_{X \ni 0}\|\Phi(X)\|\right)^n e^{nN(\Delta)}\, e^{N(V_0)} \leq C\, t_0^{-n}, \qquad (7.11)$$

where we have used that $x^n/n! \leq e^x$ and put

$$C \equiv \|A\|\, e^{N(V_0)}, \quad t_0^{-1} \equiv 2\sum_{X \ni 0}\|\Phi(X)\|\, e^{N(\Delta)}.$$

---

[99] In fact, if $\mathcal{A}_L \ni A_n \to A \in \mathcal{A}$, one has

$$\|\alpha_t^V(A) - \alpha_t(A)\| \leq \|\alpha_t^V(A - A_n)\| + \|\alpha_t^V(A_n) - \alpha_t(A_n)\|$$

$$+\|\alpha_t(A_n) - \alpha_t(A)\| \leq 2\|A - A_n\| + \|\alpha_t^V(A_n) - \alpha_t(A_n)\|,$$

and the r.h.s. can be made as small as we like. Thus, as a norm limit of elements $\alpha_t^V(A) \in \mathcal{A}$, also $\alpha_t(A) \in \mathcal{A}$.

The above estimate is enough to get the result. In fact,

$$||\alpha_t^{V_1}(A) - \alpha_t^{V_2}(A)|| \le || \sum_{n=0}^{N} (A_n^{V_1} - A_n^{V_2}) t^n ||$$

$$+ \sum_{n=N+1}^{\infty} ||A_n^{V_1} t^n|| + \sum_{n=N+1}^{\infty} ||A_n^{V_2} t^n||.$$

Now, the first term on the r.h.s. of the inequality can be made as small as we like, since, for fixed n, A_n^V becomes independent of V, for V large enough, by the finite range; moreover, by the estimate (7.11), the second and third term are smaller than $C \sum_{n=N+1}^{\infty} |t/t_0|^n$, which can be made as small as we like, for $t \le t_1 < t_0$.

The group law $\alpha_{t'} \alpha_t(A) = \alpha_{t'+t}(A)$, which is easily proved for $t, t', t+t' \in [-t_1, t-1]$, allows to extend α_t for all t. From the estimate (7.11) and the convergence of the series (7.9), it follows that α_t is strongly continuous on $\mathcal{A}_L$ and therefore also on $\mathcal{A}$.

We shall now discuss the case of an interaction potential $\Phi_1(X)$, not necessarily of finite range, satisfying the absolute summability condition

$$||\Phi_1|| \equiv \sum_{X \ni 0} ||\Phi_1(X)|| < \sum_{X \ni 0} ||\Phi(X)||, \tag{7.12}$$

with Φ of finite range, and involving only a finite number $\bar{N}$ of k-body interactions

$$\Phi_1(X) = 0, \quad \text{if } N(X) > \bar{N}. \tag{7.13}$$

For the multiple commutator $B_n(A)$, $A \in \mathcal{A}(V_0)$ (see (7.10)),

$$B_{\Phi,n}(A) = \sum_{X \subset V} [\Phi(X), B_{\Phi,n-1}(A)], \quad B_{\Phi,1}(A) = \sum_{X \subset V} [\Phi(X), A], \quad B_{\Phi,0} = A,$$

one easily proves the following algebraic identity

$$\sum_{X \subset V} [\Phi_1(X), B_{\Phi_1,n-1}(A)] - \sum_{X \subset V} [\Phi(X), B_{\Phi,n-1}(A)] =$$

$$= \sum_{m=0}^{n-1} [U_{\Phi_1}(V), ...[U_{\Phi_1-\Phi}(V), [U_{\Phi}(V), ...A]^m]]^{n-m-1}, \tag{7.14}$$

where $U_{\Phi_1}(V) \equiv \sum_{X \subset V} \Phi_1(V)$,

$$[U_{\Phi}, ...A]^m \equiv [U_{\Phi}, [U_{\Phi}, A]^{m-1}], \quad [U_{\Phi}, A]^1 = [U_{\Phi}, A].$$

Now, by applying the estimate (7.11) to the identity (7.14) we get

$$|| \sum_{X \subset V} \{[\Phi_1(X), B_{\Phi_1,n-1}(A)] - [\Phi(X), B_{\Phi,n-1}(A)]\} || \le$$

$$\leq n\,2^n ||\Phi_1 - \Phi||\,||\Phi||^{n-1}\,||A|| \prod_{m=1}^{m} [(m-1)(\bar{N}-1) + N(V_0)]$$

and therefore

$$||A_{n,1}^V - A_n^V|| \leq n\,2^n n!^{-1} ||\Phi_1 - \Phi||\,||\Phi||^{n-1}\,(N(V_0) + n(\bar{N}-1))^n\,||A||$$

$$\leq n2^n ||\Phi_1 - \Phi||\,||\Phi||^{n-1}\,e^{N(V_0)}\,e^{n\,(\bar{N}-1)}\,||A|| =$$

$$= ||A||\,||\Phi_1 - \Phi||\,||\Phi||^{-1}\,(2||\Phi||\,e^{\bar{N}-1})^n\,e^{N(V_0)} \leq C\,t_0^{-n}. \tag{7.15}$$

This estimate is enough to get the convergence of the series (7.9) for the interaction Φ_1 from that of Φ.

Another way of proving the existence of the infinite volume dynamics is to show directly that $\alpha_t{}^V(A)$ is a Cauchy sequence, i.e. for $V_2 \subset V_1$,

$$||\alpha_t{}^{V_1}(A) - \alpha_t{}^{V_2}(A)|| = ||\int_0^t ds\,(d/ds)(\alpha_s^{V_1}\alpha_{t-s}^{V_2}(A))||$$

$$= ||\int_0^t ds\,\alpha_s^{V_1}([H_{V_1} - H_{V_2},\,\alpha_{t-s}^{V_2}(A)])||$$

$$\leq \sum_{x \in V_1 \backslash V_2}\,\sum_{X \ni x}\,\int_0^{|t|} ds\,||[\Phi(X),\,\alpha_s^{V_2}(A)]||$$

converges to zero in the infinite volume limit.

This can be done for exponentially decreasing potentials by estimating $||[\alpha_t^V(A), B]||$, $A \in \mathcal{A}(\{0\})$, $B \in \mathcal{A}$.[100] For example, for two-body potentials satisfying

$$||\Phi||_\lambda \equiv \sum_{x \in Z^d} ||\Phi(\{0, x\})||\,e^{\lambda|x|} < \infty,$$

for some $\lambda > 0$, one proves that

$$||[\alpha_t^V(A), B]|| \leq ||A|| \sum_x \sup_{C \in \mathcal{A}(\{0\})} (||[\alpha_x(C), B]||/||C||)e^{-|x|\lambda + 2|t|\,||\Phi||_\lambda}$$

and the essentially finite velocity of propagation of physical disturbances

$$||[\alpha_x\alpha_t^V(A), B]|| \leq 2\,||A||\,||B||\,e^{-|t|(\lambda|x|/|t|-2||\Phi||_\lambda)}. \tag{7.16}$$

---

[100] O. Bratteli and D.W. Robinson, loc. cit. (1996), Vol. 2, Sect. 6.2.1.

8 Symmetry Breaking in Quantum Systems

Most of the wisdom on spontaneous symmetry breaking (SSB), especially for elementary particle theory, relies on approximations and/or a perturbative expansion. Since the mechanism of SSB is underlying most of the new developments in theoretical physics, it is worthwhile to try to understand it from a general (non-perturbative) point of view. Most of the popular explanations given in the literature are not satisfactory (if not misleading), since they do not make it clear that the crucial ingredient for the non-symmetrical behaviour of a system described by a symmetric Hamiltonian is the occurrence of infinite degrees of freedom and of inequivalent representations of the algebra of observables. We shall start by recalling a few basic facts.

8.1 Wigner Symmetries

The clarification of the concept of symmetry in quantum mechanics is essentially due to Wigner.[101] Given a quantum mechanical system, whose states are described by rays $\hat{\Psi} = \{e^{i\lambda}\Psi, \, \lambda \in \mathbf{R}, \, \Psi \in \mathcal{H}\}$ of a Hilbert space $\mathcal{H}$, a symmetry operation g in the sense of Wigner, briefly a *Wigner symmetry*, is a mapping of rays into rays,

$$g : \hat{\Psi} \rightarrow = g\hat{\Psi}, \tag{8.1}$$

which does not change the transition probabilities, namely the modulus of the scalar products

$$|(g\hat{\Psi}, g\hat{\Phi})| = |(\hat{\Psi}, \hat{\Phi})|.$$

As shown by Wigner,[102] any mapping satisfying the above equation can be realized either by a unitary or by an antiunitary operator $U(g)$ in $\mathcal{H}$ in the sense that

$$g\hat{\Psi} = \widehat{U(g)\Psi}. \tag{8.2}$$

$U(g)$ is determined up to a phase factor, which is irrelevant and can be eliminated by a redefinition of $U(g)$, for just one symmetry transformation.

---

[101] E.P. Wigner, *Group Theory and its Applications to the Quantum Mechanics of Atomic Spectra*, Academic Press 1959.

[102] E.P. Wigner, loc. cit.; V. Bargmann, J. Math. Phys. **5**, 862 (1964).

F. Strocchi: *Symmetry Breaking in Quantum Systems*, Lect. Notes Phys. **732**, 115–122 (2008)
DOI 10.1007/978-3-540-73593-9_8   © Springer-Verlag Berlin Heidelberg 2008

More arguments are required for a continuous group G of symmetries. If G is connected, as assumed in the sequel, the antiunitary possibility[103] is excluded, since every element is continuously connected to the identity (and can be written as the square of an element). In this case, by Wigner's theorem one has a unitary ray representation of G, namely

$$U(g)\,U(g') = \omega(g,g')\,U(gg'), \quad |\omega(g,g')| = 1 \tag{8.3}$$

and the question is whether one can select representatives $U(g)$, $g \in G$ out of the operator rays $\hat{U}(g)$, such that $\omega(g,g') = 1$. This problem has been solved by Bargmann.[104] We shall briefly sketch the argument.

First, we note that the associativity of the group multiplication, namely $((U(g)U(g'))U(g'') = U(g)(U(g')U(g''))$, implies

$$\omega(g,g')\,\omega(gg',g'') = \omega(g',g'')\,\omega(g,g'g'') \tag{8.4}$$

or, equivalently, putting $\omega(g,g') = \exp i\xi(g,g')$,

$$\xi(g,g') + \xi(gg',g'') = \xi(g',g'') + \xi(g,g'g''). \tag{8.5}$$

The analysis of the above equations is greatly simplified if, as we shall do in the sequel, we restrict the attention to *continuous* ray representations, i.e. such that $\hat{U}(g)$ is weakly continuous in g with respect to the ray scalar product.[105] This implies that one can select a strongly continuous set of representatives $U(g)$ and in this case the functions ω as well as ξ are continuous functions of the group elements.[106]

---

[103] We recall that an antiunitary operator U is antilinear, i.e. $\forall \alpha,\ \beta \in \mathbf{C},\ U\,(\alpha\Psi_1 + \beta\Psi_2) = \bar{\alpha}U\,\Psi_1 + \bar{\beta}U\,\Psi_2$, and satisfies $UU^* = U^*U = \mathbf{1}$, where the adjoint U^* is defined by $(\Psi, U^*\Phi) = \overline{(U\Psi, \Phi)},\ \forall\Psi,\ \Phi \in \mathcal{H}$. The invariance of the matrix elements under a symmetry β, i. e. $(\Psi_\beta, A_\beta\Psi_\beta) = (\Psi, A\Psi),\ \Psi_\beta \equiv U_\beta\Psi$, gives the following transformation in the antiunitary case $A_\beta = U_\beta A^* U_\beta^{-1}$, whereas in the unitary case $A_\beta = U_\beta A U_\beta^{-1}$.

[104] V. Bargmann, Ann. Math. **59**, 1 (1954).

[105] This means that

$$|(U(g)\Psi, U(g_0)\Psi)| \to |(U(g_0)\Psi, U(g_0)\Psi)| = |(\Psi, \Psi)|, \text{ if } g \to g_0.$$

[106] In fact, given a fixed unit vector Ψ, one selects $U(g)$, in a neighborhood of the identity e, by the equation $(\Psi, U(g)\Psi) \equiv |(\Psi, U(g)\Psi)|$; then, by the (ray) continuity condition $U(g)\Psi \xrightarrow{s} \Psi$ if $g \to g_0 = e$. This property extends to any Φ, by the continuity condition applied to $|(U(g)(\Psi + \lambda\Phi), U(g_0)(\Psi + \lambda\Phi))| \to (\Psi + \lambda\Phi, \Psi + \lambda\Phi),\ \lambda \in \mathbf{R}$, since, for λ sufficiently small $(\Psi, \Psi) + \mathrm{Re}\,\lambda(\Psi, \Phi) > 0$, so that also $\lambda^2\,(\Phi, U(g)\Phi) > 0$ and the convergence holds without the modulus. The extension to any g_0 follows from the unitarity of $U(g)$. For details, see Bargmann's paper quoted above and for a very elegant abstract proof see D.J. Simms, *Lie Groups and Quantum Mechanics*, Lect. Notes Math. 52, Springer 1968; Rep. Math. Phys. **2**, 283 (1971).

Then, if $g(\lambda)$, $g(\lambda')$ are two one-parameter groups in the neighborhood of the identity e, with $g(\lambda) \to e$, $g(\lambda') \to e$, when λ, $\lambda' \to 0$, we can expand all terms of (8.3) up to second order in the group parameters, e.g.

$$U(g(\lambda)) = 1 + i\lambda^a\, t_a + (1/2)\lambda^a\, \lambda^b\, t_{ab} + ..., \tag{8.6}$$

$$U(g(\lambda)\,g(\lambda')) = 1 + i(\lambda^a + \lambda'^a + \lambda^b\, \lambda'^c\, C^a_{bc})\, t_a + \tfrac{1}{2}\, (\lambda^b + \lambda'^b)(\lambda^c + \lambda'^c)t_{bc}...$$

with a, b, $c = 1, ..., N = \dim G$. Since $\omega(g, 1) = 1 = \omega(1, g)$, the expansion of $\omega(g(\lambda), g(\lambda'))$ is of the form

$$\omega(g(\lambda), g(\lambda')) = 1 + \lambda^a\, \lambda'^b\, d_{ab},$$

with d_{ab} numerical constants. Then, the comparison of the two sides of (8.3) gives

$$[t_a,\, t_b] = i f^c_{ab}\, t_c + i\, C_{ab} 1, \tag{8.7}$$

where

$$f^a_{bc} \equiv C^a_{cb} - C^a_{bc}, \quad C_{ab} \equiv d_{ba} - d_{ab}.$$

The Jacobi identity requires

$$f^a_{bc} f^e_{ad} + f^a_{cd} f^e_{ab} + f^a_{db} f^e_{ac} = 0, \tag{8.8}$$

$$f^a_{bc} C_{ad} + f^a_{cd} C_{ab} + f^a_{db} C_{ac} = 0. \tag{8.9}$$

The f's have the meaning of the structure constants of the Lie algebra L_G of G and (8.7) appears as a central extension corresponding to the Lie group $\mathcal{G}_\omega = G \times U(1)$.[107]

To quickly see when the phases can be eliminated, we note that the set of the C_{ab} defines a (real valued) antisymmetric bilinear form $C(t_a, t_b) \equiv C_{ab}$ satisfying, by (8.9),

$$C([t_a,\, t_b], t_c) + C([t_b, t_c], t_a) + C([t_c, t_a], t_b) = 0. \tag{8.10}$$

If the (simply connected) group G has the property that any bilinear form with the above properties can be written in terms of a linear form ω, in the sense that $C(t_a, t_b) = \omega([t_a, t_b])$, (technically this means that the second cohomology group $H^2(G, \mathbf{R})$ of L_G, with coefficients in $\mathbf{R}$, is trivial), then

---

[107] See Bargmann's paper and D.J. Simms' book. In fact, for the pairs (g, λ), $g \in G, \lambda \in U(1)$ the composition law

$$(g, \lambda)\,(f, \mu) = (gf, \omega(g, f)\lambda\mu)$$

satisfies associativity, thanks to (8.4), and can be shown to define a Lie group. Any continuous homomorphism $\alpha : G \to \mathcal{G}_\omega$ of the form $\alpha(g) = (g, \lambda(g))$ would satisfy $\lambda(gh) = \omega(g, h)\lambda(g)\lambda(h)$ and allow the elimination of the phases by the redefinition $\mathcal{U}(g) = \lambda(g)U(g)$.

the phases can be eliminated.[108] In fact, this is obtained by the following redefinition of the generators: $T_a = t_a + \omega(t_a)$.

8.2 Spontaneous Symmetry Breaking

The exploitation of symmetries for the description of quantum systems has played an important role in obtaining information without having to solve the full dynamical problem. It also proved useful in the case in which the symmetry is not exact by offering the possibility of unifying the description of systems related by an approximate symmetry, in terms of a "small" symmetry breaking term in the Hamiltonian in order to account for their "small" differences. Such a strategy has been successful when applied to quantum systems with a finite number of degrees of freedom, but it showed practical and conceptual difficulties when applied to infinitely extended systems.

First, the viability of such a strategy is restricted to the case of "small" symmetry breaking and therefore does not allow to unify the description of systems with rather different physical behaviour (e.g. the electromagnetic and weak interactions of elementary particles, or different thermodynamical phases in many body theory). Second, renormalization problems require an independent renormalization of the basic physical parameters, with the result of vanifying some of the possible predictions of the symmetry breaking (e.g. the electromagnetic mass differences due to isospin breaking in elementary particle theory).

From this point of view, the realization of the mechanism of *spontaneous symmetry breaking* represented a real breakthrough in the development of theoretical physics, because i) one does not have to identify a small asymmetric term in the Hamiltonian and one may use a fully symmetric Hamiltonian, ii) the symmetry breaking is accounted for by the instability of the physical world or phase chosen to describe the states of the system.

This mechanism also shows up in the classical case, where symmetric equations of motion may nevertheless lead to an asymmetric physical description, due to the existence of disjoint physical worlds or phases in which the symmetry is broken (see Part I). As we shall see also in the quantum case different phases of a system (e.g. gas, liquid and solid) with rather different physical properties can nevertheless be described by the same algebra of canonical

---

[108] The triviality of the second cohomology group allows the construction of a Lie algebra homomorphism $\alpha' : L_G \rightarrow L_{\mathcal{G}_\omega}$ of the form $\alpha'(A) = (A, \xi(A))$ (and therefore of a homomorphism $\alpha : G \rightarrow \mathcal{G}_\omega$ as discussed in the previous footnote). In fact, any linear map $\beta(A) = (A, \lambda_\beta(A)) \in L_{\mathcal{G}_\omega}$, $A \in L_G$, defines a real valued antisymmetric bilinear form $C_\beta(A, B)$

$$C_\beta(A, B) \equiv [\beta(A), \beta(B)] - \beta([A, B]),$$

which satisfies (8.10) and is therefore of the form $\omega([A, B])$. Then, $\alpha'(A) \equiv \beta(A) + \omega(A)$ yields the desired Lie algebra homomorphism.

variables and by the same dynamics, their differences being ascribed to the fact that they correspond to inequivalent representations.

A crucial role for the implementation of the above mechanism is played by the concept of *algebraic symmetry* of an algebra $\mathcal{A}$ of observables or of canonical variables, defined as an invertible mapping β of the algebra into itself, which preserves all the algebraic relations, including the $*$ (*-automorphism of $\mathcal{A}$). Clearly, if ω is a state on $\mathcal{A}$, also $\beta^*\omega$ defined by

$$(\beta^*\omega)(A) \equiv \omega(\beta(A)) \tag{8.11}$$

is a state on $\mathcal{A}$ and the corresponding GNS representations are isomorphic and physically equivalent if β commutes with the dynamics α_t. They may however yield (mathematically) inequivalent representations of $\mathcal{A}$. In this case, the corresponding vector states cannot belong to the same Hilbert space, i.e. they describe disjoint (even if equivalent) physical worlds. In a very similar way, one may introduce algebraic symmetries defined by antiautomorphisms σ of $\mathcal{A}$:

$$\sigma(\lambda A + \mu B) = \bar{\lambda}\,\sigma(A) + \bar{\mu}\,\sigma(B),$$

$$\sigma(A B) = \sigma(B)\,\sigma(A), \quad \sigma(A^*) = \sigma(A)^*, \quad \forall A, B \in \mathcal{A}.$$

They correspond to the Wigner symmetries described by antiunitary operators. For simplicity, we shall not consider this case in the sequel.

Given a representation π_ω of $\mathcal{A}$, the algebraic symmetry β gives rise to a Wigner symmetry in $\mathcal{H}_\omega$ if there exists a unitary operator U_β such that

$$U_\beta\,\pi_\omega(A)U_\beta^{-1} = \pi_\omega(\beta(A)) = \pi_{\beta^*\omega}(A). \tag{8.12}$$

The above equation is equivalent to the property that $\pi_{\beta^*\omega}$ is unitarily equivalent to π_ω. In this case, the physical description of the system in the phase $(\pi_\omega, \mathcal{H}_\omega)$ is β-symmmetric (briefly the symmetry β is unbroken or exact). On the other hand, if π_ω and $\pi_{\beta^*\omega}$ are not unitarily equivalent, there is no unitary operator U_β which implements β in $\mathcal{H}_\omega$ and the corresponding physical description is not β-symmetric. In this case the symmetry is said to be *spontaneously broken*. The name should stress the fact that one has a symmetry at the algebraic level and that the lack of symmetry of the matrix elements between states of a given representation π is due to the impossibility of describing the given algebraic symmetry by a unitary operator which maps the states of $\mathcal{H}_\pi$ into themselves.

It should be clear from the above discussion that the concept of algebraic symmetry disentangles the concept of symmetry from a concrete representation, and it is particularly useful for the description of infinite systems, for which there are generically several inequivalent representations of the algebra of observables or of canonical variables. As we shall see below, it also allows the mechanism by which a symmetry of the dynamics may fail to be a symmetry of the physical world associated to a given description of the system.

Perhaps, one of the reasons why the mechanism of spontaneous symmetry breaking has been realized so late after the foundations of quantum mechanics is that, as in the classical case, its realization crucially involves infinite degrees of freedom.

For this purpose, we consider an algebraic symmetry γ of the Weyl algebra with the property that it can be extended to the Heisenberg algebra, namely to the canonical variables q, p; from a technical point of view such a property can be formalized by the condition that γ preserves the regularity of the Weyl operators, i.e. if $\pi(U(\alpha))$, $\pi(V(\beta))$ are weakly continuous in α, β, so are $\pi(\gamma(U(\alpha)))$, $\pi(\gamma(V(\beta)))$. Indeed, by Stone's theorem such a property allows to define $\gamma(q)$, $\gamma(p)$ as the generators of the one-parameter groups $\pi(\gamma(U(\alpha)))$, $\pi(\gamma(V(\beta)))$. The algebraic symmetries of $\mathcal{A}_W$ which have this property shall be called *regular* (or *regular* *-automorphisms of $\mathcal{A}_W$).

Proposition 8.1. *If π is a regular irreducible representation of the Weyl algebra $\mathcal{A}_W$ (for finite degrees of freedom), then any regular algebraic symmetry γ of $\mathcal{A}_W$ is implemented by a unitary operator in the representation space $\mathcal{H}_\pi$* (**no spontaneous symmetry breaking**).

Proof. In fact, if π and π_γ are the GNS representations defined by the states ω and $\gamma^*\omega$ respectively, by (8.12) one has

$$(\Psi_{\gamma^*\omega}, \pi_\gamma(A)\Psi_{\gamma^*\omega}) = (\gamma^*\omega)(A) = \omega(\gamma(A)) = (\Psi_\omega, \pi(\gamma(A))\Psi_\omega).$$

Now, if ω is pure so must be $\gamma^*\omega$ since γ^* is invertible, and therefore if π is irreducible so is also π_γ. Finally, if π is regular, so is π_γ by the regularity of γ and therefore the two representations are unitarily equivalent by Von Neumann's uniqueness theorem. This means that (8.12) holds and γ is unitarily implemented.

8.3 Symmetry Breaking Order Parameter

The above characterization of spontaneous symmetry breaking as non-existence of a unitary operator implementing a given algebraic symmetry, although simple and general is not easy to check. It is therefore convenient to have a practically simpler criterium. For simplicity, we restrict our attention to the case of algebraic symmetries which commute with space and time translations, briefly called *internal symmetries*.

Proposition 8.2. *Given a representation π of the algebra $\mathcal{A}$ of observables or of canonical variables, satisfying conditions I-III of Chap. 5, and an internal symmetry β, a necessary and sufficient condition for β being unbroken in π is that all the ground state correlation functions are invariant under β, namely*

$$\omega(\beta(A)) \equiv< \beta(A) >_0 = < A >_0 = \omega(A), \quad \forall A \in \mathcal{A}, \qquad (8.13)$$

where ω denotes the ground state.

Proof. In fact, if β is unitarily implementable the state $\beta^*\omega$ (see (8.11)) is described by a vector of $\mathcal{H}_\pi$, and it is translationally invariant, since β commutes with the space translations. By the uniqueness of the translationally invariant state, it follows that $\beta^* \omega$ must coincide with ω and (8.13) follows. The converse has essentially been proved in the remark after the GNS construction at the end of Chap. 1.

A ground state expectation value $< A >_0$, such that

$$< \beta(A) >_0 \neq < A >_0,$$

will be called a *symmetry breaking order parameter*.

The above Proposition makes clear the mechanism by which a symmetry of the dynamics may nevertheless give rise to an asymmetrical physical description of the system: the point is that the states of the system are described by essentially local modifications of the ground state and states of the form $A \Psi_0$, $\beta(A) \Psi_0$ describing modifications of Ψ_0 related by the algebraic symmetry β, cannot be unitarily related if Ψ_0 is not invariant. Even if the two representations π_ω and $\pi_{\beta^* \omega}$ are physically equivalent (in the sense that they are related by a physically indistinguishable relabeling of the observables or of the coordinates $(A \to \beta(A))$), β is not a Wigner symmetry in either of them.

It is worthwhile to stress that two ingredients play a crucial role: due to the infinite number of degrees of freedom, two ground states define two disjoint worlds or phases of the system and therefore, in contrast with the case of ordinary quantum mechanics, the non-invariance of the ground state implies the asymmetry of the corresponding physical world defined by it.

The criterium of spontaneous symmetry breaking of (8.13) crucially relies on the uniqueness of the translationally invariant state and therefore it applies to pure phases. Symmetric correlation functions defined by a mixed state do not imply that the symmetry is unbroken in the pure phases in which the theory (defined by such correlation functions) decomposes. The check of the symmetry of the correlation functions should then be accompanied by the check of the cluster property.

The criterium of Proposition 8.2 also holds if β is only assumed to commute with the time translations, π is irreducible and the uniqueness of the translationally invariant state is replaced by the uniqueness of the ground state, i.e. if π satisfies conditions I, II, III i) of Chap. 5 (but not necessarily III ii).

In fact, $[\beta, \alpha_t] = 0$ implies that $V(\beta, t) \equiv U_\beta U(t) U_\beta^{-1} U(-t)$ commutes with $\mathcal{A}$ and therefore, by the irreducibility of π, is a multiple of the identity, say $\exp i\, h(\beta, t)\, \mathbf{1}$ with h a real function. The strong continuity of $U(t)$ and the group law imply that h is a continuous function of t and actually a linear function $h(\beta, t) = t\, h(\beta)$, i.e.

$$U_\beta U(t) U_\beta^* = e^{it\, h(\beta)}\, U(t), \quad U_\beta^* U(t) U_\beta = e^{-it\, h(\beta)}\, U(t).$$

The above equations are incompatible with the energy spectral condition unless $h = 0$, since

$$U(t) U_\beta^n \Psi_0 = e^{-int\, h(\beta)} U_\beta^n \Psi_0, \quad U(t) U_\beta^{*n} \Psi_0 = e^{int\, h(\beta)} U_\beta^{*n} \Psi_0.$$

Thus, $U_\beta \Psi_0$ is invariant under $U(t)$ and by the uniqueness of the ground state, it must be of the form $\exp(i\alpha)\Psi_0$, $\alpha \in \mathbf{R}$. Equation (8.13) then follows easily.

9 Examples

To illustrate the above general ideas we discuss simple concrete models exhibiting spontaneous symmetry breaking.

1. Heisenberg Ferromagnet

The Heisenberg model for spin $1/2$ Ferromagnets is described by the following finite lattice Hamiltonian

$$H_V = - \sum_{i,j \in V} J_{ij} \, \boldsymbol{\sigma}_i \cdot \boldsymbol{\sigma}_j - \mathbf{h} \cdot \sum_{j \in V} \boldsymbol{\sigma}_j, \qquad (9.1)$$

where V denotes the finite three dimensional lattice, $\mathbf{h}$ is an external uniform magnetic field, i, j label the lattice points and J_{ij} is the positive coupling constant or "potential", invariant under lattice translations and of short range, e.g. a nearest neighbor coupling (see Sect. 7.1).

   As discussed in Sect. 7.1, the algebraic dynamics α_t is defined as the norm limit of the finite volume dynamics α_t^V generated by H_V. The spin rotations define a three parameter group of *-automorphisms or algebraic symmetries of the quasi local spin algebra $\mathcal{A}$, which commute with the time translations α_t in the limit $h = |\mathbf{h}| \to 0$.

   For finite V, the ground state $\Psi_0{}^{V,h}$ (defined on $\mathcal{A}_V$ and by Hahn-Banach extension on $\mathcal{A}$) is characterized by all the spins pointing in the direction of $\mathbf{n} \equiv \mathbf{h}/|\mathbf{h}|$, i.e.

$$\boldsymbol{\sigma}_j \cdot \mathbf{n} \, \Psi_0{}^{V,h} = \Psi_0{}^{V,h}.$$

The correlation functions of $\Psi_0{}^{V,h}$ converge as $V \to \infty$ and define a state Ω_0^h on $\mathcal{A}$, which is invariant under space translations and under α_t. In fact, thanks to the uniform convergence of α_t^V, one has

$$\Omega_0^h(\alpha_t(A)) \equiv \lim_{V' \to \infty} \lim_{V \to \infty} \Omega_0^{V,h}(\alpha_t^{V'}(A))$$

$$= \lim_{V \to \infty} \Omega_0^{V,h}(\alpha_t^V(A)) = \lim_{V \to \infty} \Omega_0^{V,h}(A) = \Omega_0^h(A).$$

Moreover, by keeping $\mathbf{n}$ fixed and letting $h \to 0$, the correlation functions of Ω_0^h converge and define a state $\Omega_0^{\mathbf{n}}$ on $\mathcal{A}$, which is not invariant under spin

F. Strocchi: *Examples*, Lect. Notes Phys. **732**, 123–125 (2008)
DOI 10.1007/978-3-540-73593-9_9   © Springer-Verlag Berlin Heidelberg 2008

rotations. This gives rise to a symmetry breaking order parameter (*magnetization*)

$$\Omega_0^{\mathbf{n}}(\mathbf{n} \cdot \boldsymbol{\sigma}_j) = \lim_{h \to 0} \lim_{V \to \infty} (\Psi_0^{V,h}, \mathbf{n} \cdot \boldsymbol{\sigma}_j \Psi_0^{V,h}) = 1. \qquad (9.2)$$

Each $\Omega_0^{\mathbf{n}}$ defines a (physically relevant) representation $\pi_{\mathbf{n}}$ of $\mathcal{A}$ with cyclic vector $\Psi_0^{\mathbf{n}}$. Different directions $\mathbf{n}$ give rise to inequivalent representations of $\mathcal{A}$, each labeled by a different symmetry breaking order parameter. In fact, by asymptotic abelianess, the ergodic limits

$$\lim_{V \to \infty} V^{-1} \sum_{i \in V} \mathbf{n} \cdot \boldsymbol{\sigma}_i \equiv (\mathbf{n} \cdot \boldsymbol{\sigma})_\infty$$

exist in any $\pi_{\mathbf{n}}$ and belong to the center (by the proof of Proposition 6.3); since they take different values in representations labeled by different $\mathbf{n}$, such representations cannot be unitarily related. Such representations are physically equivalent in the sense that one goes from one to the other by a different choice of the coordinate axes (which leaves the Hamiltonian invariant); the physically relevant point is the existence of a symmetry breaking order parameter in each $\pi_{\mathbf{n}}$.

By taking rotationally invariant averages of the states $\Omega_0^{\mathbf{n}}$, similarly to (7.6) for the free Bose gas, one may obtain a state Ω_0^{inv} whose correlation functions are rotationally invariant and do not provide a symmetry breaking order parameter. However, Ω_0^{inv} is not a pure state on $\mathcal{A}$ and symmetry breaking order parameters emerge if one decomposes Ω_0^{inv} into the pure states of which is a mixture.

2. Bose-Einstein Condensation

As discussed in Sect. 7.2, the gauge transformations define a one-parameter group of algebraic symmetries of the field algebra $\mathcal{A}$, which describes a system of free bosons.

In each representation π_θ defined by Ω_θ, the gauge symmetry is spontaneously broken with order parameter $< \psi >_\theta$. The occurrence of symmetry breaking also for a free system is due to the fact that for non-zero density the total number operator does not exist, the generalized version of Von Neumann's theorem does not apply and inequivalent representations of the Weyl field algebra are allowed.

On the other hand, all the correlation functions of the gauge invariant state Ω are by construction invariant under gauge transformations and one may wonder about their breaking. The point is that Ω is a pure state on the observable algebra but not on the field algebra $\mathcal{A}$, as displayed by the violation of the cluster property by the correlation functions of $\mathcal{A}$. The symmetry breaking emerges when one makes a decomposition into the pure states of which Ω is a mixture.

The model is an interesting example of the mechanism of spontaneous breaking of a gauge symmetry, which by definition reduces to the identity on the observables.

3. Massless Field in $d \geq 3$

For the free massless field in space time dimensions $d \geq 3$ (see Chap. 2; for $d = 2$ the model is infrared singular[109]), the "gauge" transformations $\beta^\lambda(\varphi(x)) = \varphi(x) + \lambda$ define internal algebraic symmetries, which are spontaneously broken in any physically relevant representation, with symmetry breaking order parameter $< \varphi >_0 \neq < \beta^\lambda(\varphi) >_0$.

---

[109] See e.g. F. Strocchi, *Selected Topics on the General Properties of Quantum Field Theory*, World Scientific 1993, Sect. 7.2.

10 Constructive Symmetry Breaking

Apart from simple models, like those discussed in the previous section, the existence of a symmetry breaking order parameter is a non-trivial problem which in principle requires the control on the correlation functions. In this section we briefly discuss constructive criteria for symmetry breaking.

A. Goldstone (Perturbative) Criterium

In a pioneering paper on symmetry breaking, Goldstone discussed a quantum (scalar) field theory model exhibiting a symmetry breaking order parameter.[110] Since the standard perturbative expansion based on the standard Fock representation predicts the vanishing of the field expectation value, Goldstone suggested a strategy which combines a perturbative expansion and a semiclassical approximation (*Goldstone criterium*). Since this has become the standard approach to symmetry breaking within the perturbative approach, it is worthwhile to discuss briefly the Goldstone criterium.

The Goldstone model is described by the following Lagrangean

$$\mathcal{L} = \tfrac{1}{2} \partial_\mu \varphi \, \partial^\mu \varphi - U(\varphi), \quad U(\varphi) = \lambda(\varphi^2 - a^2)^2, \tag{10.1}$$

where φ is a real scalar field transforming as an n-dimensional irreducible representation of the internal symmetry group $O(n)$. The Goldstone strategy is based on the following steps:

i) (*semiclassical approximation*) one considers the classical absolute minima φ_{min} of the (classical) potential U (which form an orbit under $O(n)$)

ii) (*perturbative expansion about the mean field semiclassical approximation*) one picks up one absolute minimum φ_{min} and builds up a perturbative quantum expansion around such a classical value of the field: $\varphi = \varphi_{min} + \chi$.

The expansion is conveniently organized as a quantum (or loop) expansion in $\hbar$.[111] It is an important result that such an expansion makes sense, namely

---

[110] J. Goldstone, Nuovo Cim. **10**, 154 (1961). For a general critical discussion and for an outline of the constructive approach to symmetry breaking in quantum field theory, see A.S. Wightman, Constructive Field Theory, in *Fundamental Interactions in Physics and Astrophysics*, (Coral Gables 1972), G. Iverson et al. eds., Plenum 1973.

[111] S. Coleman and E. Weinberg, Phys. Rev. **D7**, 1888 (1973).

F. Strocchi: *Constructive Symmetry Breaking*, Lect. Notes Phys. **732**, 127–130 (2008)
DOI 10.1007/978-3-540-73593-9_10   © Springer-Verlag Berlin Heidelberg 2008

that a renormalized perturbation expansion exists.[112] It then follows that in such a perturbative expansion

$$< \varphi >_0 = \varphi_{min} + \text{small quantum corrections}$$

and therefore, if φ_{min} is not symmetric, so is $< \varphi >_0$. In this way one constructs a (perturbative) theory with a symmetry breaking order parameter. By this logic, each absolute minimum identifies a ground state and a nonsymmetric theory.

B. Ruelle Non-perturbative Strategy

The Goldstone strategy has proved successful for application to many body theory (e.g. the Ginzburg-Landau model of superconductivity) and for elementary particle theory (see the perturbative treatment of the standard model of electromagnetic and weak interactions), but it leaves some basic questions open. In fact, it is known that mean field approximations are often not reliable and the results on the triviality of the φ^4 theory in four space time dimensions seem to indicate that the perturbative expansion, which might be, at best, an asymptotic expansion, may have little to do with the non-perturbative solution.

A strategy for a non-perturbative approach to symmetry breaking in quantum field theory and in many body theory is provided by the imaginary time (or euclidean) formulation and the functional integral representation of the euclidean correlation functions.[113]

For the Goldstone model this is obtained by introducing a space cutoff V (e.g. by working in a finite volume V) and an ultraviolet cutoff K (e.g. by replacing the continuous euclidean space by a regular lattice). Then, the imaginary time correlation functions are given by a functional integral

$$< \varphi(x_1)...\varphi(x_n) >_{V,K} = Z_{V,K}^{-1} \int \mathcal{D}\varphi \, e^{- \int_V \mathcal{L}_{ren}(\varphi_K) \, dx} \varphi_K(x_1)...\varphi_K(x_n),$$

(10.2)

where φ_K denotes the (euclidean) field on the (finite) lattice, with lattice spacing $a = K^{-1}$, and $\mathcal{L}_{ren}$ the renormalized euclidean Lagrangean, (including the infrared and ultraviolet counterterms needed to ensure the convergence of the correlation functions, when the cutoffs are removed, according

---

[112] B.W. Lee, Nucl. Phys. **B9**, 649 (1969); K. Symanzik, Renormalization of Theories with Broken Symmetry, in *Cargèse Lectures in Physics 1970*, D. Bessis ed., Gordon and Breach, New York 1972; C. Becchi, A. Rouet and R. Stora, Renormalizable Theories with Symmetry Breaking, in *Field Theory, Quantization and Statistical Physics*, E. Tirapegui ed., D. Reidel 1981. For textbook accounts, see e.g. J. Collins, *Renormalization*, Cambridge Univ. Press 1984, Chap. 9; L.S. Brown, *Quantum Field Theory*, Cambridge Univ. Press 1994.

[113] See J. Glimm and A. Jaffe, *Quantum Physics. A Functional Integral Point of View*, 2nd ed., Springer 1987. For a handy account see [SNS 96].

to the non-perturbative renormalization mentioned in Chap. 3) and

$$Z_{V,K} = \int \mathcal{D}\varphi \, e^{-\int_V \mathcal{L}_{ren}(\varphi_K) \, dx}. \tag{10.3}$$

In this way, the problem takes the form of a problem of statistical mechanics, with Z playing the role of the partition function, and one may use the well established strategy for the existence of a symmetry breaking order parameter in statistical systems.

This strategy has been discussed at length with mathematical rigor in Ruelle's book[114], and we shall briefly call it the *Ruelle strategy*. The general idea is to compute the above correlation functions with specified boundary conditions for φ_K, e.g. $\varphi_K = \overline{\varphi}$ on the boundary ∂V, and discuss the dependence of the thermodynamical limit $(V \to \infty)$ on the boundary conditions. It is a deep result that, under general conditions, any state can be obtained in this way by a suitable choice of the boundary conditions, and, therefore, if the thermodynamical limit of the correlation functions is independent of the boundary conditions (as it happens above the critical temperature), there is only one phase and no spontaneous symmetry breaking.

On the other hand, the dependence on the boundary conditions indicates that there is more than one phase and if different boundary conditions, related by a symmetry operation, give rise to different correlation functions (in the thermodynamical limit and when $K \to \infty$), then there is symmetry breaking.

In fact, if g is an internal symmetry (therefore leaving the Lagrangean invariant) and one chooses as boundary condition $\varphi_K = \overline{\varphi}$ on ∂V, one has, putting $\varphi^g \equiv g\,\varphi$,

$$< \varphi^g(x_1)...\varphi^g(x_n) >_{V,K,\overline{\varphi}} =$$

$$Z_{V,K,\overline{\varphi}}^{-1} \int \mathcal{D}\varphi \, e^{-(A_V(\varphi)+A_{\partial V}(\overline{\varphi}))} \varphi_K^g(x_1)...\varphi_K^g(x_n), \tag{10.4}$$

where A_V denotes the euclidean (renormalized) action and $A_{\partial V}$ the boundary term which enforces the chosen boundary condition.

Now, since the Lagrangean, and therefore the action, is invariant under the symmetry g, by a change of variables in the functional integral, say $\varphi_K' \equiv \varphi_K^g$, the right hand side of the above equation becomes

$$Z_{V,K,\overline{\varphi}}^{-1} \int \mathcal{D}\varphi' \, e^{-(A_V(\varphi')+A_{\partial V}(g^{-1}\overline{\varphi}'))} \varphi_K'(x_1)...\varphi_K'(x_n) =$$

$$< \varphi(x_1)...\varphi(x_n) >_{V,K,g^{-1}\overline{\varphi}}. \tag{10.5}$$

Thus, the non-invariance of the above correlation functions in the thermodynamical limit is equivalent to the dependence on the (non-symmetric) boundary conditions.

---

[114] D. Ruelle, *Statistical Mechanics*, Benjamin 1969. For the applications see also G.L. Sewell, *Quantum Theory of Collective Phenomena*, Oxford Univ. Press 1986, esp. Part III, and B. Simon, *The Statistical Mechanics of Lattice Gases*, Vol. I, Princeton Univ. Press 1993.

Clearly, if the chosen boundary conditions are symmetric (e.g. periodic boundary conditions), the corresponding correlation functions are invariant, but this cannot be taken as a criterium for absence of spontaneous symmetry breaking, because the so constructed correlation functions may correspond to a mixed phase or to a representation with more than one translationally invariant state as displayed by the failure of the cluster property.

C. Bogoliubov Strategy

Another constructive way of obtaining symmetry breaking order parameters was discussed by Bogoliubov[115] and exploited in particular in his treatment of superconductivity. The idea is to introduce a symmetry breaking interaction with an external field, which is sent to zero at the very end. Such a prescription looks more physical, since it reflects the operational way of producing, e.g., a ferromagnet, but does not seem to be under the same rigorous mathematical control as is the Ruelle strategy.

In the Goldstone model discussed above, the idea of the Bogoliubov strategy can be implemented by introducing in the (infrared and ultraviolet) regularized theory an n-component external field $h(x)$ which plays the role of the external magnetic field for ferromagnets, linearly coupled to $\varphi(x)$ (more generally one may modify the coupling constant). Clearly, the volume interaction with the external field wins over the surface terms due to the boundary conditions and the latter ones become irrelevant.

Then, one computes the correlation functions in the thermodynamical limit and *finally* one lets $h \to 0$. Proceeding as in the above discussion of the Ruelle strategy, one easily gets the following relation between the infinite volume correlation functions:

$$< \varphi^g(x_1)...\varphi^g(x_m) >_{K,n} = < \varphi(x_1)...\varphi(x_m) >_{K,g^{-1}n}, \qquad (10.6)$$

where n denotes the direction along which h is sent to zero.

The criterium of symmetry breaking associated with the Bogoliubov strategy is then the following: if in the thermodynamical limit the so obtained correlation functions do not depend on the direction along which $h \to 0$, one has only one phase and no symmetry breaking. On the other hand, if different directions of h give rise to different limits, one obtains non-invariant correlation functions and spontaneous symmetry breaking. Indeed, the Bogoliubov procedure closely corresponds to the way by which one operationally produces a non-trivial magnetization in a given direction.

The non-uniqueness of the thermodynamical limit, in the strategies discussed above, is an indication of a sort of dynamical instability, since an infinitesimally small interaction (a surface or boundary term or a vanishingly small volume interaction with an external field) is capable of drastically changing the state in the thermodynamical limit and the physical behaviour of the system.

---

[115] N.N. Bogoliubov, *Lectures on Quantum Statistics*, Vol. 2, Gordon and Breach 1970, Part 1.

11 Symmetry Breaking in the Ising Model

Most of the theoretical wisdom on the phase transition of the ferromagnetic type and the related symmetry breaking is based on the two-dimensional Ising model, which also played the role of a laboratory for ideas and strategies and it is now regarded as a corner stone in the foundations of statistical mechanics. Anyone interested in critical phenomena and in the functional integral approach to quantum field theory should have a look at the model. Even if a discussion of the two-dimensional Ising model would be very appropriate for our purposes, we refer the reader to the very good accounts which can be found in literature[116]. We restrict our discussion to the one-dimensional version of the model, which is almost trivial, but nevertheless provides an interesting simple example for testing the constructive strategies of symmetry breaking discussed above.

The Ising model was invented to mimic the phenomenon of ferromagnetism and it is a simplified version of the Heisenberg model. The algebra $\mathcal{A}$, which describes the degrees of freedom of the system, is the spin algebra generated by polynomials of the spins in various sites (see Sect. 7.1), and the finite volume Hamiltonian is

$$H_V = -J \sum_{i \in V} \sigma_i \, \sigma_{i+1} - h \sum_{i \in V} \sigma_i, \qquad (11.1)$$

where, for the spin $1/2$ case, $s_i \equiv \sigma_i/2$ denotes the $z-$ component of the spin at the i-th lattice site. The inversion of the spins $\gamma(\sigma_i) = -\sigma_i$ is an internal symmetry and we shall see that it is spontaneously broken at zero temperature.[117]

---

[116] For the history of the model, see S.G. Brush, Rev. Mod. Phys. **39**, 883 (1967). The model is now part of the basic knowledge in statistical mechanics and the theory of phase transitions; for textbook accounts, see e.g. K. Huang, *Statistical Mechanics*, Wiley 1987, Chap. 14, 15; G. Gallavotti, *Statistical Mechanics: A Short Treatise*, Springer 1999, Sect. 6; B. Simon, *The Statistical Mechanics of Lattice Gases*, Vol. I, Princeton Univ. Press 1993, Sect. II.6. An extensive treatment, which also emphasizes the links with quantum field theory and general theoretical physics problems, is in B.M. McCoy and T.T. Wu, *The Two Dimensional Ising Model*, Harvard Univ. Press 1973.

[117] For the basic elements of statistical mechanics, see e.g. K. Huang, *Statistical Mechanics*, Wiley 1987; a brief account is given in the following section.

F. Strocchi: *Symmetry Breaking in the Ising Model*, Lect. Notes Phys. **732**, 131–138 (2008)
DOI 10.1007/978-3-540-73593-9_11   © Springer-Verlag Berlin Heidelberg 2008

The model can also be used to describe a lattice gas, with the choice $n_i \equiv (1 + \sigma_i)/2 = 1$ if the i-th site is occupied by a "molecule", and $n_i = 0$ otherwise. The Hamiltonian models an interaction between the "molecules" by a square well potential of the form $U(r) = \infty$ for $r < a \equiv$ the lattice spacing, $U(r) = -U$ for $a < r < 2a$, and $U(r) = 0$ for $r > 2a$. In this interpretation J is related to U and $h = 2\mu + d$, where μ is the chemical potential and d is the number of nearest neighbors per site; the fluid phase would correspond to $< n_i >= 1$ and the gas to $< n_i >= 0$.

The calculation of the spin correlation functions at non-zero temperature $T = 1/\beta$, in the thermodynamical limit, is a very simple but instructive example of how the general wisdom of statistical mechanics works in concrete examples and as such can be regarded as a prototype of the functional integral approach to quantum field theory models. We shall discuss the model in the case of a one-dimensional lattice, with the purpose of illustrating the various strategies of constructive symmetry breaking, discussed in the previous section.

1. Free Boundary Conditions

We consider the case $h = 0$ with free boundary conditions (i.e. no boundary condition) in finite volume, i.e. for N sites.

The partition function is

$$Z_N = \sum_{\sigma_1 = \pm 1} \cdots \sum_{\sigma_N = \pm 1} e^{\beta J \sum_{i=1}^{N-1} \sigma_i \sigma_{i+1}} \qquad (11.2)$$

and can be easily computed by noting that, since σ_i takes only the values ± 1 and cosh is an even function,

$$\sum_{\sigma_N = \pm 1} e^{\beta J \sigma_{N-1} \sigma_N} = 2 \cosh(\beta J \sigma_{N-1}) = 2 \cosh \beta J.$$

Thus, a recursive application of the argument gives

$$Z_N = 2^N \left(\cosh \beta J \right)^{N-1}.$$

All the correlation functions are γ symmetric (as in (10.5)). In fact, by a change of variables $(\sigma \to \sigma' = -\sigma)$ one has, $\forall \beta$,

$$< \sigma_{k_1} \ldots \sigma_{k_n} >_N = Z_N^{-1} \sum_\sigma \sigma_{k_1} \ldots \sigma_{k_n} e^{\beta J \sum_{i=1}^{N-1} \sigma_i \sigma_{i+1}} =$$

$$Z_N^{-1} \sum (-1)^n \sigma'_{k_1} \ldots \sigma'_{k_n} e^{\beta J \sum_{i=1}^{N-1} \sigma'_i \sigma'_{i+1}} = (-1)^n < \sigma_{k_1} \ldots \sigma_{k_n} >_N.$$

Thus, all the correlation functions of an odd number of spins vanish, i.e. all the correlation functions are symmetric. To say something on symmetry breaking, one has to control what happens in the pure phases, i.e. one must check the cluster property in the thermodynamical limit.

For this purpose, we compute the two point function which can be easily obtained by the following trick: we modify the model by introducing site dependent couplings J_i and introduce the corresponding partition function

$$Z_N(J_i) = 2^N \prod_{i=1}^{N-1} \cosh(J_i\beta).$$

Then, one has

$$Z_N < \sigma_k \, \sigma_{k+r} >_N = \left(\sum_\sigma \sigma_k \sigma_{k+r} \, e^{\beta \sum_i J_i \sigma_i \, \sigma_{i+1}} \right)_{J_i=J} =$$

$$\beta^{-r} \left(\frac{\partial}{\partial J_k} \frac{\partial}{\partial J_{k+1}} \cdots \frac{\partial}{\partial J_{k+r-1}} Z_N(J_i) \right)_{J_i=J} = Z_N \, (\tanh \beta J)^r. \qquad (11.3)$$

This formula displays the independence of $< \sigma_k \, \sigma_{k+r} >_N$ from the number of lattice sites so that it coincides with its thermodynamical limit and shows the invariance under lattice translations. Quite generally, for ordered sites one gets

$$< \sigma_k \sigma_{k+r_1} \, \sigma_j \sigma_{j+r_2} \cdots \sigma_l \, \sigma_{l+r_n} >_N = (\tanh(\beta J))^{\sum_{i=1}^n r_i}.$$

Then, one has $\forall \beta < \infty$

$$\lim_{r \to \infty} < \sigma_k \, \sigma_{k+r} > = 0,$$

whereas in the limit $\beta \to \infty$

$$< \sigma_k \sigma_{k+r} >^{T=0} = 1.$$

Thus, the cluster property fails at zero temperature, which means that the correlation functions computed with free boundary conditions define a mixed state (at $T = 0$). This teaches us the general lesson that in the presence of symmetry breaking the thermodynamical limit taken without any boundary condition leads to a violation of the cluster property.

By the same argument as above, one can show that all the correlation functions at $T \neq 0$ satisfy the cluster property and therefore their symmetry proves that there is no spontaneous symmetry breaking at non-zero temperature.

2. Periodic and Cyclic Boundary Conditions

A commonly used choice is that of periodic boundary conditions, mainly because they have the virtue of preserving translational invariance in finite volume. But, being invariant under internal symmetries, they also lead to a mixed phase, when there is symmetry breaking.

The computation of the correlation functions with periodic boundary conditions is instructive also because it allows the use of the *transfer matrix*, which has become a powerful tool in statistical mechanics and in lattice quantum field theory.[118] To this purpose, the exponential

$$T(i, i+1) \equiv e^{\beta J \sigma_i \sigma_{i+1} + \beta h (\sigma_i + \sigma_{i+1})/2} = T(i+1, i) \tag{11.4}$$

can be viewed as the matrix element $< \sigma_i |T| \sigma_{i+1} >$ of an operator T, called the *transfer matrix*, between vectors $|\sigma_i >$ labeled (only) by the value (± 1) taken by the spin σ_i, e.g. $|\sigma_i >= |+ >= |\sigma_{i+1} >$, if $\sigma_i = 1 = \sigma_{i+1}$. Thus, T is effectively acting on a two dimensional space and is given by

$$T = \begin{pmatrix} T_{++} & T_{+-} \\ T_{-+} & T_{--} \end{pmatrix} = \begin{pmatrix} e^{\beta J + \beta h} & e^{-\beta J} \\ e^{-\beta J} & e^{\beta J - \beta h} \end{pmatrix}. \tag{11.5}$$

Its eigenvalues are

$$\lambda_{\pm}(h) = e^{\beta J} \cosh \beta h \pm [e^{2\beta J} \sinh^2(\beta h) + e^{-2\beta J}]^{1/2}, \tag{11.6}$$

Then, the partition function becomes

$$Z_N = \sum_{\sigma_1, \sigma_N} < \sigma_1 |T^{N-1}| \sigma_N > e^{\beta h (\sigma_1 + \sigma_N)/2}. \tag{11.7}$$

Z_N is easily computed for periodic boundary conditions, $\sigma_1 = \sigma_N$, if $h = 0$, since it is given by the trace of T^{N-1}

$$Z_N = \lambda_+^{N-1} + \lambda_-^{N-1}, \quad \lambda_+ = 2 \cosh \beta J, \quad \lambda_- = 2 \sinh \beta J. \tag{11.8}$$

The correlation functions can be computed with the trick of introducing site dependent couplings, as before, and by taking derivatives of

$$Z_N(J_i) = \prod_{i=1}^{N-1} \lambda_+^i + \prod_{i=1}^{N-1} \lambda_-^i,$$

since the $T(J_i)$ are all simultaneously diagonalizable.

For $\beta < \infty$, the thermodynamical limit is dominated by the highest eigenvalue $\lambda_+ > \lambda_-$, for N large

$$Z_N = \lambda_+^{N-1} (1 + (\lambda_-/\lambda_+)^{N-1}) \sim \lambda_+^{N-1}.$$

In this limit one gets the same results as for the case of free boundary conditions, as expected.

The partition function can be easily computed also if one imposes *cyclic boundary conditions*, by which the open line of the lattice is turned into a

---

[118] See T.D. Schulz, D.C. Mattis and E.H. Lieb, Rev. Mod. Phys. **36**, 856 (1964) and references therein; E. Lieb, in *Boulder Lectures in Theoretical Physics*, Vol.XI D, K.T. Mahantappa and W.E. Brittin eds., Gordon and Breach 1969, p.329; J.B. Kogut, Rev. Mod. Phys. **51**, 659 (1979).

circle with the identification $\sigma_{N+1} \equiv \sigma_1$. Then, the Hamiltonian reads

$$H_N = -J \sum_{i=1}^{N} \sigma_i \, \sigma_{i+1} - h \sum_{i=1}^{N} \sigma_i \qquad (11.9)$$

and one has

$$Z_N = \text{Tr} \left(T(1,2)...T(N, N+1) \right) = \sum_{\sigma_1} < \sigma_1 | \, T^N \, | \sigma_1 > = \lambda_+(h)^N + \lambda_-(h)^N.$$

Also in this case, for $h = 0$, one gets symmetric correlation functions and a violation of the cluster property at $T = 0$.

3. Ruelle Strategy. Symmetry Breaking Boundary Conditions

According to the general discussion of the previous section, the pure phases can be obtained by an appropriate choice of the boundary conditions, in this case by symmetry breaking boundary conditions.

In fact, for boundary conditions $\sigma_1 = \sigma_N = \sigma_B$ and for $h = 0$, one has (by (11.7))

$$Z_N = < \sigma_B | T^{N-1} | \sigma_B > .$$

We start with the case $\beta < \infty$ (non-zero temperature). In this case, the transfer matrix T has strictly positive entries and by the Perron-Frobenius theorem, the largest eigenvalue λ_+ is non-degenerate.[119] Hence, if $|\lambda_+ >$ denotes the eigenstate with the highest eigenvalue and P the corresponding projection, one has for large N, if $< \sigma_B | \lambda_+ > \neq 0$, $(\sqrt{2}|\lambda_+ > = |+> + |->)$

$$Z_N \sim \lambda_+^{N-1} < \sigma_B | P | \sigma_B > .$$

To compute the (average) magnetization $< \sigma >$, we consider a spin chain of $2N + 1$ sites, centered at the origin; then, for large N,

$$< \sigma_0 >_{2N+1} = Z_{2N+1}^{-1} \sum_{\sigma_0 = \pm} < \sigma_B | T^N | \sigma_0 > \sigma_0 < \sigma_0 | T^N | \sigma_B > \sim$$

$$\sim < \sigma_B | P | \sigma_B >^{-1} < \sigma_B | P \, \tau_3 \, P | \sigma_B > = 0,$$

where we have used that

$$\lim_{N \to \infty} T^N / \lambda_+^N = P,$$

[119] For the proof of this result, and its relevance in the functional integral approach to quantum theories, see J. Glimm and A. Jaffe, *Quantum Physics. A Functional Integral Point of View*, 2nd ed., Springer 1987, p. 51.

and that, in terms of the spin Pauli matrices τ_i, one has

$$P = (1 + \tau_1)/2, \quad \sum_{\sigma_0} |\sigma_0 > \sigma_0 < \sigma_0| = \tau_3, \quad P\tau_3 P = 0.$$

On the other hand, for $\beta \to \infty$, T is no longer strictly positive, actually

$$T = e^{\beta J} \mathbf{1}, \quad \lambda_+ = \lambda_- \equiv \lambda, \quad Z_N = \lambda^{N-1}$$

and

$$< \sigma_0 >_{2N+1} = < \sigma_B | \tau_3 | \sigma_B > = \pm 1, \quad \text{if } \sigma_B = \pm 1.$$

Thus, at zero temperature the magnetization equals the spin value at the boundary. By the same technique, one may compute, e.g. the two point function and check that the cluster property is satisfied. In conclusion, with Ruelle's strategy one gets pure phases and symmetry breaking at zero temperature.

4. Bogoliubov Strategy

It is not difficult to check Bogoliubov strategy in this model, by working with a non-zero magnetic field. In this case, the thermodynamical limit is independent of the boundary conditions and the computation is particularly simple if one uses cyclic boundary conditions.

The magnetization is obtained by taking the derivative of Z_N with respect to βh and one gets in the thermodynamical limit

$$< \sigma_k > = \frac{e^{\beta J} \sinh(\beta h)}{[e^{2\beta J} \sinh^2(\beta h) + e^{-2\beta J}]^{1/2}}. \tag{11.10}$$

Now, for any non-zero temperature (i.e. $\beta < \infty$), the limit $h \to 0$ vanishes independently of the direction of h.

By the same trick, one may prove that all correlation functions have a limit independent of the direction along which $h \to 0$ and therefore by Bogoliubov criterium, there is only one phase and no symmetry breaking.

On the other hand, for $T = 0$ (i.e. $\beta \to \infty$), one has

$$< \sigma_k >_h^{T=0} = h/|h|. \tag{11.11}$$

Thus, the limit $h \to 0^\pm$ depends on the direction of h and there are two possible values of the magnetization, corresponding to two different phases. In each phase there is symmetry breaking.

5. Mean Field Approximation

Finally, it is worthwhile to check how the mean field approximation, which is related to the *Goldstone criterium*, compares with the exact solution.

The approximation is defined by expanding the spin configurations on the lattice around a mean magnetization $< \sigma >$, to be determined at the end

self-consistently[120], and by keeping only the lowest order terms. This leads
to the following finite volume Hamiltonian

$$H_V^{mean} = -2J\sum_{i\in V}\sigma_i < \sigma > -h\sum_{i\in V}\sigma_i = -(2J < \sigma > +h)\sum_{i\in V}\sigma_i, \quad (11.12)$$

where the factor 2 accounts for the number of nearest neighbors for each site.

The corresponding partition function is the same as that of a non-interacting chain of spins in the presence of an (effective) external field $h_{eff} = h + 2J < \sigma >$ and it is easily computed:

$$Z_N = 2^N(\cosh \beta h_{eff})^N.$$

Thus, the one point function in the limit $h \to 0$ is given by

$$< \sigma_k >= Z_N^{-1}N^{-1}\partial Z_N/\partial(\beta h)|_{h=0} = \tanh(2\beta J < \sigma >).$$

This formula has a trivial solution for the magnetization, $< \sigma >= 0$, but also a non-trivial solution whenever $T < T_c \equiv 2J$. Thus, the mean field approximation predicts spontaneous symmetry breaking also for non-zero temperature, in disagreement with the exact solution. The point is that, for $T \neq 0$, the fluctuations induced by the neglected terms $O(s^2)$, in the expansion $\sigma =< \sigma > +s$, win over the lowest order terms and wash out the order parameter.

It is worthwhile to mention that, quite generally, the mean field approximation has the following structural features:

i) it replaces the original symmetric Hamiltonian (with zero external field) by a non-symmetric one and actually leads to a description of the system based on a dynamics which depends on the order parameter; in the exact treatment instead, as stressed before, the dynamical law is the same in all phases and is therefore independent of the order parameter (only the correlation function are). In a certain sense, the mean field mixes algebraic properties with properties related to the ground state.
ii) it replaces a short range dynamics, e.g. corresponding to a nearest neighbor coupling, by an infinite range dynamics, since the average spin $< \sigma >$ coincides with the expectation of

$$\sigma_\infty \equiv \lim_{V\to\infty} V^{-1}\sum_{i\in V}\sigma_i$$

---

[120] This approximation is at the basis of the Curie-Weiss theory of magnetic phase transitions, also called *molecular field approximation*; see e.g. H.E. Stanley, *Introduction to Phase Transitions and Critical Phenomena*, Oxford Univ. Press 1974, Chap. 6; C.J. Thompson, *Mathematical Statistical Mechanics*, Princeton Univ. Press 1972, Sect. 4.5.

(see Chap. 6), which involves all the spins. In a certain sense, the mean field approximation mimics a long range dynamics and, in fact, it shares some of the basic features of long range interactions leading to long range delocalization, as it occurs in Coulomb systems and in gauge theories (in positive gauges, see Chap. 19). For a discussion of such common features, which play a crucial role for the energy spectrum of the Goldstone theorem, see G. Morchio and F. Strocchi, Erice Lectures 1985, in *Fundamental Problems of Gauge Field Theory*, G. Velo and A.S. Wightman eds., Plenum 1986).

12 * Thermal States

The physically relevant representations discussed in Chap. 5 are characterized by the existence of a lowest energy or ground state and are supposed to describe states of an infinitely extended isolated system. The situation changes if one wants to describe states of a system at non-zero temperature (*thermal states*), i.e. states of a system in thermal equilibrium with a reservoir. The stability of the system is now guaranteed by the reservoir and there is no need of the energy spectral condition. The role of the ground state is now taken by the *equilibrium state* and one is therefore led to discuss representations of the canonical or observable algebra defined by equilibrium states.

As for the zero temperature case, one expects substantial differences with respect to the finite dimensional case; for infinitely extended systems, the Gibbs factor becomes meaningless in general, because the formal (Fock) Hamiltonian becomes ill defined in the infinite volume limit. The strategy is to extract from the finite dimensional case those properties which survive the thermodynamical limit.[121]

As a first step we shall discuss the characterization of the equilibrium states, Sects. 12.1–12.3; then we shall identify those states which describe pure phases, Sect. 12.4.

12.1 Gibbs States and KMS Condition

According to the principles of quantum statistical mechanics,[122] the equilibrium states of a system in a finite volume V are described by density matrices. For the description of a system in terms of *Gibbs canonical ensemble* (fixed number of particles), the equilibrium states are given by the following expectations, for any bounded operator A,

$$\Omega_\beta(A) = Z_\beta^{-1} \operatorname{Tr}(\rho_\beta A), \quad \rho_\beta = e^{-\beta H}, \quad Z_\beta = \operatorname{Tr} e^{-\beta H}, \qquad (12.1)$$

---

[121] Here we give a sketchy account, in view of the discussion of symmetry breaking at non-zero temperature. For a beautiful and more detailed presentation, see R. Haag, *Local Quantum Physics*, 2nd ed., Springer 1996, Chap.V and H.M. Hugenholtz, in *Mathematics of Contemporary Physics*, R.F. Streater ed., Academic Press 1972.

[122] See e.g. P.A.M. Dirac, *The Principles of Quantum Mechanics*, 4th ed., Claredon Press Oxford 1958, Sect. 33; K. Huang, loc. cit. 1987.

F. Strocchi: * *Thermal States*, Lect. Notes Phys. **732**, 139–150 (2008)
DOI 10.1007/978-3-540-73593-9_12   © Springer-Verlag Berlin Heidelberg 2008

where $\beta = 1/T$ is the inverse temperature, H is the Hamiltonian and Tr denotes the trace in the Hilbert space $\mathcal{H}$ of the states of the system in the volume V at the given temperature.

Similarly, in the case of *Gibbs grand canonical ensemble*, corresponding to a description in which the number of particles is not fixed, the equilibrium states are given by

$$\Omega_{\beta,\mu}(A) = Z_{\beta,\mu}^{-1} \,\mathrm{Tr}\,(\rho_{\beta,\mu}\, A), \quad \rho_{\beta,\mu} = e^{-\beta(H-\mu N)}, \quad Z_{\beta,\mu} = \mathrm{Tr}\,\rho_{\beta,\mu}, \quad (12.2)$$

where μ is the chemical potential and N is the number operator.

For simplicity, we shall often drop the subscripts β and μ and we shall generically refer to the states defined by (12.1), (12.2) as *Gibbs states* on the C^*-algebra $\mathcal{A}_V = \mathcal{B}(\mathcal{H})$ of all bounded operators in $\mathcal{H}$.

For the thermodynamical limit, the use of the grand canonical ensemble is more suitable and we shall in general consider the corresponding states; for simplicity, sometimes we shall still denote by H the "grand canonical Hamiltonian" $H(\mu) \equiv H - \mu N$.

It follows easily from (12.1), (12.2) that the Gibbs states are invariant under time evolution, i.e. they are equilibrium states, e.g.

$$\Omega_\beta(\alpha_t(A)) = Z_\beta^{-1}\,\mathrm{Tr}(\rho_\beta\, e^{iHt}\, A\, e^{-iHt}) = \Omega_\beta(A),$$

since ρ_β commutes with H.

The states defined by (12.1), (12.2) are not pure states (see (1.7)) and therefore the GNS representations defined by them are not irreducible. Furthermore, for non-zero particle density, the average energy, which is non-zero at non-zero temperature, diverges in the infinite volume limit, in agreement with the physical expectation that the energy per particle is non-zero in the limit. Thus, the definition of the Hamiltonian in the thermodynamical limit becomes problematic and suitable subtractions are needed. One is therefore facing the basic problem of the description of an infinite system and of the mathematical status of the thermodynamical limit at non-zero temperature. Clearly, the framework discussed in Chap. 4 for the zero temperature case requires substantial changes.

For this purpose, we recall a few basic mathematical properties of the Gibbs states. First, we recall that for a system of free particles in a box $\exp(-\beta H_0)$, where H_0 is the free Hamiltonian, is of trace class[123], i.e. $\mathrm{Tr}\,|\exp(-\beta H_0)| < \infty$. Under general conditions on the interaction potential[124] also $\exp(-\beta H)$ is of trace class, for all positive β's.

---

[123] For the properties of trace class operators, see e.g. M. Reed and B. Simon, *Methods of Modern Mathematical Physics*, Vol. I, Academic Press 1972, Sect. VI.6.

[124] D. Ruelle, Helv. Phys. Acta **36**, 789 (1963); J. Lebowitz and E. Lieb, Adv. Math. **9**, 316 (1972), Appendix by B. Simon. A sufficient condition is that the potential U is a small perturbation, i.e. that for any $a < 1$ there is a $b \geq 0$ such that $|(\Psi,\, U\,\Psi)| < a\,(\Psi,\, H_0\,\Psi) + b\,(\Psi,\Psi)$, for all Ψ in the domain of the free Hamiltonian H_0. In fact, the above inequality implies $H \geq (1-a)\,H_0 - b\mathbf{1}$ and $e^{-\beta H_0}$ of trace class implies $e^{-\beta H}$ of trace class.

Since the product of a bounded operator and a trace class operator is an operator of trace class[125], also $\exp\left(-\beta\left(H - \mu N\right)\right)$ is of trace class for all β, for μ in a suitable range, so that $H - \mu N > 0$. Thus, under such general conditions (12.1), (12.2) are well defined.

When dealing with systems in a finite volume, we shall always assume that ρ_β and/or $\rho_{\beta,\mu}$ is of trace class.

Since the thermodynamical limit is a convenient extrapolation for the description of very large systems, it is physically reasonable to try to extract those structural properties of the finite dimensional case which are expected to be stable in the limit.

Theorem 12.1. [126] *Under the above general conditions a Gibbs state, given by (12.1) or (12.2), satisfies the* **KMS-condition,**[127] *namely*
$$\forall A,\, B \in \mathcal{B}(\mathcal{H})$$

$$F^{\beta}_{AB}(t) \equiv \Omega_\beta(B\,\alpha_t(A)), \quad G^{\beta}_{AB}(t) \equiv \Omega_\beta(\alpha_t(A)\,B) \qquad (12.3)$$

are boundary values of analytic functions $F^{\beta}_{AB}(z)$, $G^{\beta}_{AB}(z)$, analytic in the strips $0 < Im\,z < \beta$ and $-\beta < Im\,z < 0$, respectively, and

$$F^{\beta}_{AB}(t + i\beta) = G^{\beta}_{AB}(t). \qquad (12.4)$$

Conversely, any state satisfying the KMS-condition, briefly called a **KMS-state,** *for all bounded operators in a Hilbert space $\mathcal{H}$ is a Gibbs state.*

---

[125] See e.g. M. Reed and B. Simon, *Methods of Modern Mathematical Physics*, Vol. I, Academic Press, Theorem VI.19. The point is that the operators of trace class form a vector space. In fact, for any partial isometry S, by using Schwarz' inequality one has

$$|\mathrm{Tr}\,(S|A|)| \leq \sum_n \||A|^{1/2}\,S^*\Psi_n\|\,\||A|^{1/2}\,\Psi_n\|$$

$$\leq \left(\sum_n \||A|^{1/2}S^*\Psi_n\|^2\right)^{1/2} \left(\sum_n \||A|^{1/2}\,\Psi_n\|^2\right)^{1/2} = \mathrm{Tr}(S|A|S^*)^{1/2}\,(\mathrm{Tr}\,|A|)^{1/2}$$

and $\mathrm{Tr}\,(S|A|S^*) \leq \mathrm{Tr}\,|A|$. Then, if U, U_A, U_B denote the partial isometries occurring in the polar decomposition of $A + B$, A, B, respectively,

$$\mathrm{Tr}\,|A + B| \leq |\mathrm{Tr}\,(U^*U_A|A|)| + |\mathrm{Tr}\,(U^*U_B|B|)| \leq \mathrm{Tr}\,|A| + \mathrm{Tr}\,|B|.$$

Hence, in order to prove that $A\,B$ is of trace class if B is so and A is bounded, it suffices to consider the case in which A is self-adjoint and of norm less than one; in this case A is a linear combination of the unitary operators $U_\pm \equiv A \pm i(1 - A^2)^{1/2}$, and $U_\pm B$ is clearly of trace class, if B is so, since $|U_\pm B| = |B|$.

[126] R. Haag, N.M. Hugenholtz and M. Winnink, Comm. Math. Phys. **5**, 215 (1967), hereafter referred as [HHW].

[127] R. Kubo, J. Phys. Soc. Jap. **12**, 570 (1957); P.C. Martin and J. Schwinger, Phys. Rev. **115**, 1342 (1959).

Proof.   Given a bounded operator A, for any $0 \le \gamma \le \beta$

$$A_{t+i\gamma} \, e^{-\beta H} \equiv e^{-\gamma H} \, e^{iHt} \, A e^{-iHt} \, e^{-H(\beta-\gamma)}$$

is a bounded operator of trace class, since it is the product of bounded operators with at least one of trace class; furthermore, for $0 < \gamma < \beta$, is differentiable in t, γ (since for any $\delta > 0$, $He^{-\delta H}$ is a bounded operator) and satisfies the Cauchy-Riemann equations, so that

$$F^{\beta}_{AB}(z) \equiv \mathrm{Tr}\,(e^{-\beta H} \, B A_z) = \mathrm{Tr}\,(B A_z \, e^{-\beta H}) = \mathrm{Tr}\,(A_z \, e^{-\beta H} \, B) \qquad (12.5)$$

is an analytic function of $z = t + i\gamma$, for $0 < \mathrm{Im}\, z < \beta$. Similarly one proves the analyticity of $G^{\beta}_{AB}(z)$ for $-\beta < \mathrm{Im}\, z < 0$. The KMS boundary condition, (12.4), follows by taking the boundary value of (12.5) at $\mathrm{Im}\, z = \beta$.

Conversely, if for given β the KMS condition holds for the state Ω, then Ω is invariant under time translations, since (12.4) for $B = 1$, $A = A^*$ gives

$$F^{\beta}_A(t + i\beta) = G^{\beta}_A(t) = F^{\beta}_A(t),$$

i.e. $F^{\beta}_A(t)$ is periodic in the direction of the imaginary axis, hence analytic and bounded in the whole complex plane. Hence $F^{\beta}_A(t)$ is a constant.

Now, a state Ω on $\mathcal{B}(\mathcal{H})$ can be written as $\mathrm{Tr}(\rho_\Omega A)$ with ρ_Ω a positive matrix of trace equal to one and the KMS condition for $t = 0$ gives

$$\mathrm{Tr}(\rho_\Omega B e^{-\beta H} \, A e^{\beta H}) = \mathrm{Tr}\,(\rho_\Omega \, A B), \quad \forall B \in \mathcal{B}(\mathcal{H}).$$

This implies

$$e^{-\beta H} \, A e^{\beta H} \rho_\Omega = \rho_\Omega A,$$

i.e. $[e^{\beta H} \rho_\Omega, \, A] = 0$, $\forall A \in \mathcal{B}(\mathcal{H})$. Hence,

$$\rho_\Omega = Z^{-1} \, e^{-\beta H}, \quad Z = \mathrm{Tr}\, e^{-\beta H}.$$

An equivalent form of the KMS condition, which will turn useful in the applications is the following.[128] Let $\tilde{F}^{\beta}(w)$, $\tilde{G}^{\beta}(w)$ be the (distributional) Fourier transforms of $F^{\beta}(t)$, $G^{\beta}(t)$, (defined in (12.3)), respectively. Then

$$\tilde{F}^{\beta}(w) = e^{-\beta w} \, \tilde{G}^{\beta}(w). \qquad (12.6)$$

This follows from the fact that $F^{\beta}(t)$, $G^{\beta}(t)$ and $F^{\beta}(t + i\beta)$ are all bounded continuous functions of t, hence tempered distributions, and the Fourier transform of $F^{\beta}(t + i\beta)$ is $e^{\beta w} \tilde{F}^{\beta}(w)$. By a similar argument, one shows that the KMS condition is equivalent to the following one

$$\int f(t - i\beta) \, F^{\beta}(t) dt = \int f(t) \, G^{\beta}(t) dt, \quad \forall \tilde{f} \in \mathcal{D}(\mathbf{R}). \qquad (12.7)$$

---

[128] [HHW].

12.2 GNS Representation Defined by a Gibbs State

As we shall see below, the main virtue of the KMS condition with respect to the Gibbs formula ((12.1) or (12.2)) is that the former one survives the thermodynamical limit and can therefore be used to characterize the equilibrium states in this limit, whereas the Gibbs formula becomes meaningless.

The physical reason is that in the infinite volume limit the average energy diverges. On the other hand, by Theorem 12.1, quite generally the KMS condition implies the invariance of the state under time translations and therefore the existence of a one-parameter group of (strongly continuous) unitary operators $U(t)$, implementing the time translations, in the GNS representation defined by such a KMS state. The apparent conflict between the existence of the generator of $U(t)$ and the divergence of the average energy requires a better understanding of the structure of the GNS representation defined by a KMS state. Again, we shall start from the case of finite volume.

From the definition of a Gibbs state, we have that $Z_\beta^{-1} \rho_\beta$ is a positive operator and we denote by r_0 its square root; it is a Hilbert-Schmidt operator, i.e. such that $r_0^* r_0$ is of trace class. The vector space D_0 of Hilbert-Schmidt operators is invariant under right and left multiplication by bounded operators[129] and it is naturally equipped by a Hilbert scalar product

$$(\Psi_r, \Psi_r) \equiv \mathrm{Tr}\,(r^* r), \tag{12.8}$$

where Ψ_r denotes the vector identified by the Hilbert-Schmidt operator r. Actually, D_0 is a Hilbert space, i.e. it is closed under the topology τ defined by the above scalar product.[130] Then, the Gibbs state defined by ρ_β (or by $\rho_{\beta,\mu}$) can be written as

$$\Omega_\beta(A) = \mathrm{Tr}\,(r_0 A r_0) = (\Psi_{r_0}, \Psi_{A r_0}) \equiv (\Psi_{r_0}, \pi(A)\,\Psi_{r_0}). \tag{12.9}$$

The above equation is well defined since $A r_0 = 0$ implies $A = 0$: in fact, as an operator in the GNS representation space $\mathcal{H}_\beta$, $r_0^{-1} = Z_\beta^{1/2}\,e^{\beta H/2}$ has a dense domain D and therefore $0 = A r_0(r_0^{-1} D) = A D$ implies $A = 0$. This shows that Ψ_{r_0} is a *separating vector* for the algebra $\mathcal{A}_V$, in this subsection simply denoted by $\mathcal{A}$.

---

[129] See e.g. M. Reed and B. Simon, loc. cit., Theorem VI.22.

[130] In fact, τ convergence implies operator norm convergence, since

$$\|B\|^2 = \|B^* B\| = \sup_{\|x\|=1}\,(x, B^* B x) \leq \mathrm{Tr}\,(B^* B).$$

Furthermore, the convergence of $\mathrm{Tr}\,(B_n^* B_n)$ implies that the sequence is uniformly bounded, so that $\sum_{k=1}^N \|B_n x_k\|^2 \leq C$ uniformly in n, N and $\sum_{k=1}^N \|B x_k\|^2 \leq C$, uniformly in N; hence, the limit operator B has a finite Hilbert-Schmidt norm.

Furthermore, Ψ_{r_0} is a cyclic vector for $\mathcal{A}$, since

$$(\Psi_r, \pi(\mathcal{A}) \Psi_{r_0}) = \text{Tr}\,(r^* \mathcal{A} r_0) = 0$$

implies $\text{Tr}\,(r^* r r_0 r_0) = \text{Tr}\,(r_0 r^* r r_0) = 0$, i.e. $r\,r_0 = 0$ and therefore $r = 0$.

In conclusion, (12.9) displays the explicit GNS representation π of $\mathcal{A}$ as operators in the GNS representation space $\mathcal{H}_\beta = D_0$, with a cyclic and separating vector Ψ_{r_0}.

The so-constructed GNS representation space D_0 is also the carrier of a conjugate, i.e. antilinear, representation π' of $\mathcal{A}$ given by

$$\pi'(A)\Psi_{r_0} = \Psi_{r_0 A^*}. \tag{12.10}$$

Clearly, $\pi'(\lambda A) = \bar{\lambda}\,\pi'(A)$, $\forall \lambda \in \mathbf{C}$, where $\bar{\lambda}$ denotes the complex conjugate of λ. Furthermore,

$$||\pi(A)|| = ||\pi'(A)|| = ||A||, \tag{12.11}$$

so that the representation is *faithful*.[131]

In order to characterize the infinite volume limit of the KMS states, we need to derive other general properties of the GNS representation defined by a Gibbs state (which we shall show to be stable under the thermodynamical limit). Such additional information is provided by the following theorem.

Theorem 12.2. [132] *The GNS representation defined by a Gibbs state has the following properties*
i) the commutant $\pi(\mathcal{A})'$ of $\pi(\mathcal{A})$ is the weak closure of $\pi'(\mathcal{A})$

$$\pi(\mathcal{A})' = (\pi'(\mathcal{A}))'', \tag{12.12}$$

(equivalently $\pi(\mathcal{A})'' = (\pi'(\mathcal{A}))'$)
ii) there exists an antiunitary operator J such that

$$J\,\pi(\mathcal{A})\,J = \pi'(\mathcal{A}), \quad \forall A \in \mathcal{A}, \tag{12.13}$$

$$J^2 = \mathbf{1}, \tag{12.14}$$

$$J\Psi_{r_0} = \Psi_{r_0}. \tag{12.15}$$

---

[131] In fact,

$$||\pi(A)||^2 = \sup_r \text{Tr}\,(r^* A^* A r)/\text{Tr}\,(r^* r) = \sup_r \text{Tr}\,(A r r^* A^*)/\text{Tr}\,(r^* r)$$

$$= \sup_r \text{Tr}\,(A r^* r A^*)/\text{Tr}(r r^*) = ||\pi'(A)||^2.$$

The equality $||\pi(A)||^2 = ||A||^2$ follows since one can choose r as the projection on a state with spectral support relative to $A^* A$ as close as one likes to $||A||^2$.

[132] [HHW]; see also the book by Haag (1996) and the London lectures by Hugenholtz (1972), quoted in footnote 121.

Proof.

i) By definition

$$\pi(A)\,\pi'(B)\,\Psi_{r_0} = \Psi_{A r_0 B^*} = \pi'(B)\pi(A)\Psi_{r_0},$$

so that $\pi(\mathcal{A}) \subseteq (\pi'(\mathcal{A}))'$. By taking the weak closure one gets

$$\pi(\mathcal{A})'' \subseteq ((\pi'(\mathcal{A}))'')' = (\pi'(\mathcal{A}))'. \tag{12.16}$$

On the other hand, the subalgebra $\mathcal{A}_0 \subseteq \mathcal{A}$ of Hilbert-Schmidt operators is a Hilbert algebra[133] and for such algebras

$$\pi(\mathcal{A}_0)'' = (\pi'(\mathcal{A}_0))'. \tag{12.17}$$

Then,

$$\pi(\mathcal{A})'' \supseteq \pi(\mathcal{A}_0)'' = (\pi'(\mathcal{A}_0))' \supseteq \pi'(\mathcal{A})'. \tag{12.18}$$

Equations (12.16), (12.18) imply (12.12).

ii) The operator J is defined by $J\Psi_r = \Psi_{r^*}$. Equations (12.14), (12.15) are obvious and (12.13) follows from

$$J\,\pi(A)J\,\Psi_r = J\,\pi(A)\Psi_{r^*} = J\Psi_{Ar^*} = \Psi_{rA^*} = \pi'(A)\Psi_r. \tag{12.19}$$

We can now clarify the relation between the generator of the time translations and the many particle (Fock) Hamiltonian.

The time invariance of a Gibbs state Ω_β implies that in the corresponding GNS representation the time translations are implemented by a one-parameter group of unitary operators $U(t)$, $t \in \mathbf{R}$ (we omit the finite volume suffix V). The weak continuity of α_t implies that $U(t)$ can be chosen weakly and therefore strongly continuous. Now, the condition

$$U(t)\,\pi(A)\,U(t)^{-1} = \pi(\alpha_t(A)) \tag{12.20}$$

implies

$$U(t) = \pi(U_F(t))\,V(t), \tag{12.21}$$

where $U_F(t)$ is generated by the Fock Hamiltonian H_F (or by the Fock operator $H_F - \mu N$) (for simplicity, the two possibilities will both be denoted by $\tilde{H}_F$) and $V(t) \in \pi(\mathcal{A})'$.

The invariance of Ψ_{r_0} under $U(t)$ uniquely fixes $V(t)$. In fact, since $\pi(\mathcal{A})' = \pi'(\mathcal{A})''$, $V(t)$ is of the form $V(t) = \pi'(V_F(t))$, $V_F(t) \in \mathcal{B}(\mathcal{H}_F) = \mathcal{A}$ and therefore

$$\Psi_{r_0} = \pi(U_F(t))\,\pi'(V_F(t))\,\Psi_{r_0} = \Psi_{U_F(t)r_0\,V_F(t)^*}.$$

Since r_0 commutes with $U_F(t)$ and the representation is faithful

$$r_0(1 - U_F(t)\,V_F(t)^*) = 0, \quad \text{i.e.} \quad V_F(t) = U_F(t).$$

---

[133] J. Dixmier, *Von Neumann algebras*, North-Holland 1981, Chap. 1.

In conclusion the generator of $U(t)$ is not the Fock Hamiltonian H_F (or $H_F - \mu N$) but

$$H = \pi(\tilde{H}_F) - \pi'(\tilde{H}_F) \tag{12.22}$$

where the second term on the right hand side has the meaning of the contribution to the energy by the reservoir.

It is instructive to work out the case of a (quantum) lattice spin system and explicitly check the properties i), ii) (in particular (12.17) is easily proven). We shall leave this exercise to the reader[134].

As we shall see below, the occurrence of a subtraction in the definition of the generator of the time translations allows the existence of the generator H also in the infinite volume limit, when the average energy becomes divergent and the Fock Hamiltonian H_F (or $H_F - \mu N$) does not exist.

12.3 KMS States in the Thermodynamical Limit

The power of the KMS condition is that it makes sense also in the thermodynamical limit and can be used as a characterization of the equilibrium states in such a limit, where the Gibbs prescription become meaningless. It can actually be proven that the KMS condition survives the thermodynamical limit under general conditions, as stated in Theorem 12.3 below.

For this purpose, a few comments on the thermodynamical limit are useful. A state Ω_V describing the system in a finite volume V is a positive linear functional on $\mathcal{A}(V) \subseteq \mathcal{A}$, ($\mathcal{A} \equiv$ the quasi local algebra, see Chap. 4). By the Hahn-Banach theorem it can be extended to $\mathcal{A}$ (the extension of the state will still be denoted by Ω_V) and therefore, as V varies, one gets a sequence $\{\Omega_V\}$ of states on $\mathcal{A}$. Since the closed unit ball of the dual $\mathcal{A}^*$ of a Banach space $\mathcal{A}$ is compact in the weak topology induced by $\mathcal{A}$, (Alaoglu-Banach theorem), then by the Bolzano-Weierstrass theorem there is a subsequence Ω_{V_n} which is weakly convergent, i.e. $\forall A \in \mathcal{A}$ one has the existence of

$$\lim_{n \to \infty} \Omega_{V_n}(A).$$

In conclusion, by the above compactness argument, one can always find a thermodynamical limit of finite volume states. In general, such a limit will not be unique; different limits correspond to different boundary conditions leading to different phases.

Theorem 12.3. [135] *Let α_t^V denote the finite volume (algebraic) dynamics defined by $U_V(t) = e^{iH_V t} \in \mathcal{A}$, (where H_V denotes the finite volume Hamiltonian), and Ω_V denote the KMS (Gibbs) states at inverse temperature β (and with given chemical potential μ).*

*If α_t^V converges in norm as $V \to \infty$ on the quasi local algebra $\mathcal{A}$ to a one-parameter group α_t of * automorphisms of $\mathcal{A}$ and Ω is the weak limit of*

---

[134] For help see Hugenholtz's lectures in London (1972).
[135] [HHW].

finite volume states on $\mathcal{A}$

$$\Omega(A) = \lim_{n \to \infty} \Omega_{V_n}(A), \quad \forall A \in \mathcal{A}, \tag{12.23}$$

then Ω satisfies the KMS condition.

Proof. In order to prove the KMS condition in the form (12.7), it is enough to prove that

$$\lim_{V \to \infty} \Omega_V(B\,\alpha_t^V(A)) = \Omega(B\,\alpha_t(A)), \quad \forall B,\, A \in \mathcal{A}.$$

Now, putting $A_t^V \equiv \alpha_t^V(A)$ we have

$$|\Omega_V(B\,A_t^V) - \Omega(BA_t)| \le |\Omega_V(B(A_t^V - A_t))| + |(\Omega_V - \Omega)(BA_t)|.$$

Since Ω_V is a continuous functional on $\mathcal{A}$, the first term on the right hand side is bounded by $\|B\|\,\|A_t^V - A_t\|$, which goes to zero as $V \to \infty$ by assumption, and the second term converges to zero if Ω is the weak limit of Ω_V.

One can also show[136] that the general properties of KMS (Gibbs) states derived in Theorem 12.2 remain valid in the thermodynamical limit, namely the representation π_β defined by a KMS state Ω_β has the following properties:

1) there exists an involution operator J, with $J^2 = \mathbf{1}$, such that

$$J\,\pi(\mathcal{A})''\, J = \pi(\mathcal{A})', \quad J\Psi_\beta = \Psi_\beta, \tag{12.24}$$

   where the vector Ψ_β represents Ω_β in $\mathcal{H}_\beta$
2) the time translations are implemented by strongly continuous unitary operators $U(t)$, such that

$$U(t)\,\Psi_\beta = \Psi_\beta, \quad [U(t),\, J] = 0, \tag{12.25}$$

3) the generator H of $U(t)$ satisfies

$$e^{-\beta H/2}\,\pi(A)\,\Psi_\beta = J\,\pi(A)^*\,\Psi_\beta, \quad \forall A \in \mathcal{A} \tag{12.26}$$

   and by (12.25) $HJ - JH = 0$.

12.4 Pure Phases. Extremal and Primary KMS States

In this subsection we shall discuss the characterization of the pure phases in thermodynamics.

First we recall that the thermodynamical phases are defined by equilibrium states and the above discussion indicates that the KMS states, being

---

[136] [HHW]; see also Haag's book (1996) and Hugenholtz's lectures in London (1972), where one can also find the connection with the Tomita-Takesaki theory of Von Neumann algebras.

the thermodynamical limit of equilibrium Gibbs states, are the natural candidates for describing equilibrium states[137]. Thus, one has to identify the property of KMS states which corresponds to the phase being pure.

In the zero temperature case, the pure phases were identified by irreducible representations, but now, by the previous discussion, in particular (12.24), the GNS representation defined by a KMS state is not irreducible and actually its commutant is as big as the weak closure of $\pi(\mathcal{A})$. The physical interpretation of such a violation of irreducibility is that the role of the commutant is to account for the "degrees of freedom" of the reservoir, whose interaction with the system is needed in order to keep the temperature constant. Thus, irreducibility cannot be used to characterize the pure phases at non-zero temperature.

The relevant property is that the concept of pure phase is related to that of equilibrium state which is not a mixture of other equilibrium states. KMS states which cannot be decomposed as mixture of other KMS states are called *extremal* and therefore the pure thermodynamical phases can be described by extremal KMS states.

In general, since in the standard thermodynamical sense pure phases are defined by homogeneous states, one adds the condition that the corresponding KMS states are invariant under space translations.

In conclusion, with respect to the zero temperature case discussed in Chap. 5, for non-zero temperature the conditions which selects the physically relevant representations have to be modified as follows.

I. (**Existence of energy and momentum**) The space and time translations are implemented by strongly continuous groups of unitary operators (as in Chap. 5).

II. (**Thermodynamical stability**) The representation is defined by a KMS state.

III. (**Equilibrium state**) The KMS state is the unique translationally invariant state.

The **pure phases** are defined by extremal KMS states.

For extremal KMS states, condition III can be replaced by the validity of the cluster property, as in Chap. 6, thanks to Proposition 6.4, which states such an equivalence for factorial representations. Whereas in the zero temperature case factoriality was implied by irreducibility, in the non-zero temperature case the equivalence between extremal and factorial KMS representations is given by the following theorem.

Theorem 12.4. *A KMS state Ω is extremal iff its GNS representation π is factorial, i.e. the center $\mathcal{Z} = \pi(\mathcal{A})'' \cap \pi(\mathcal{A})'$ consists of multiples of the identity.*

---

[137] For further arguments involving stability properties, see Haag's book (1996), Sect. V.3.

Proof. [138] The proof is split into four steps.

1) A KMS state Ω is extremal iff there is no other KMS state ω_1, which is not a multiple of Ω, such that

$$\omega_1 \leq \lambda \Omega, \quad \lambda > 1. \tag{12.27}$$

In fact, if ω_1 exists one has the decomposition

$$\Omega = \lambda^{-1} \omega_1 + (\Omega - \lambda^{-1} \omega_1) \equiv \omega_1' + \omega_2.$$

Conversely, if Ω is decomposable in terms of KMS states as $\Omega = \omega_1 + \omega_2$, then clearly there exists $\omega_1 < \Omega$.

2) Equation (12.27) implies (the vector Ψ_Ω represents Ω)

$$\omega_1(B^* A)^2 \leq \omega_1(B^* B)\,\omega_1(A^* A) \leq \lambda^2\,\Omega(B^* B)\,\Omega(A^* A) =$$

$$\lambda^2 ||\pi(B)\Psi_\Omega||^2\,||\pi(A)\,\Psi_\Omega||^2, \quad \forall A, B \in \mathcal{A}.$$

Thus, $\omega_1(B^* A)$ defines a bounded (densely defined) sesquilinear form on the GNS representation space defined by Ω and therefore there exists a unique bounded operator T' such that

$$\omega_1(B^* A) = (\pi(B)\Psi_\Omega,\, T'\,\pi(A)\Psi_\Omega). \tag{12.28}$$

Furthermore, $T' \in \pi(\mathcal{A})'$ since $\forall A, B, C, \in \mathcal{A}$

$$(\pi(B)\Psi_\Omega,\, T'\,\pi(C)\,\pi(A)\,\Psi_\Omega) = \omega_1(B^* C A) = \omega_1((C^* B)^* A) =$$

$$(\pi(C^* B)\Psi_\Omega,\, T'\pi(A)\,\Psi_\Omega) = (\pi(B)\Psi_\Omega,\, \pi(C)T'\,\pi(A)\Psi_\Omega).$$

3) T' is invariant under time translations

$$T_t' \equiv U(t)\,T'\,U(t)^{-1} = T'. \tag{12.29}$$

In fact, $[T_t',\, A] = U(t)\,[T',\, A_{-t}]U(-t) = 0, \forall A \in \mathcal{A}$ implies $T_t' \in \pi(\mathcal{A})'$ and then

$$(\pi(A)\,\Psi_\Omega,\, T_t'\,\pi(B)\,\Psi_\Omega) = (\Psi_\Omega,\, \pi(A)^*\,\pi(B)\,T_t'\Psi_\Omega) =$$

$$(\Psi_\Omega,\, (\pi(A^* B))_{-t}\,T'\Psi_\Omega) = \omega_1((A^* B)_{-t}) = \omega_1(A^* B) =$$

$$= (\pi(A)\,\Psi_\Omega,\, T'\,\pi(B)\,\Psi_\Omega),$$

where the invariance of Ω and ω_1 under time translations has been used.

---

[138] Here we give a brief sketch; for a detailed proof, see e.g. O. Bratteli and D.W. Robinson, *Operator Algebras and Quantum Statistical Mechanics*, Vol. II, Springer 1981, Proposition 5.3.29.

4) Define
$$T \equiv J T' J. \tag{12.30}$$

It belongs to $\pi(\mathcal{A})''$ by (12.24) and also to $\pi(\mathcal{A})'$. In fact, the time translation invariance of Ψ_Ω and (12.26) give

$$T \Psi_\Omega = J T' J \Psi_\Omega = J T' \Psi_\Omega = J T' e^{-\beta H/2} \Psi_\Omega =$$

$$= J e^{-\beta H/2} T' \Psi_\Omega = T' \Psi_\Omega,$$

where in the last step we have used that $T' \in \pi(\mathcal{A})'$ and the strong closure of (2.26). Therefore

$$\omega_1(A) = (\Psi_\Omega, T' A \Psi_\Omega) = (\Psi_\Omega, A T' \Psi_\Omega) = (\Psi_\Omega, A T \Psi_\Omega) = \Omega(A T).$$

To conclude the argument, we use the KMS condition in the following form

$$\omega(A B_t) = \omega(B_{t-i\beta} A),$$

which can be easily derived in the same way as (12.5), and the above relation $\Omega(A T) = \omega_1(A)$; thus, we get

$$\Omega(ATBC) = \Omega(\alpha_{-i\beta}(BC)AT) = \omega_1(\alpha_{-i\beta}(BC)A) =$$

$$\omega_1(\alpha_{-i\beta}(B)\alpha_{-i\beta}(C)A) = \omega_1(\alpha_{-i\beta}(C)AB) =$$

$$= \Omega(\alpha_{-i\beta}(C)ABT) = \Omega(ABTC),$$

i.e. $[T, B] = 0$, $\forall B \in \mathcal{A}$. In conclusion, since $T \in \mathcal{Z}$, Ω is extremal iff $\mathcal{Z} = \{\lambda \mathbf{1}, \lambda \in \mathbf{C}\}$, i.e. its GNS representation is factorial, briefly iff Ω is a *factor state*.

The physical relevance of the concept of factor, also called *primary*, state is that in the GNS representation defined by it macroscopic observables like ergodic means or variables at infinity have a sharp (classical) value in agreement with the physical picture of a pure phase in thermodynamics.

13 Fermi and Bose Gas at Non-zero Temperature

As an example of symmetry breaking at non-zero temperature we discuss the free Fermi and Bose gas, starting from finite volume and then discussing the thermodynamical limit.

1. Free Fermi and Bose Gas in Finite Volume

We consider a system of free fermions or bosons in a finite volume V with a free Hamiltonian H_0 defined by periodic boundary conditions (for simplicity, for the moment we omit the label V which denotes that we are in a finite volume). [139] One can then use a Fock representation and the non-zero temperature states are the Gibbs states. In view of the thermodynamical limit to be considered later, it is convenient to use grand canonical states Ω, (12.2) with the chemical potential μ to be fixed in such a way that the average density $\Omega(N)/V$ takes a given value $\bar{\rho}$.

Since $H(\mu) = H_0 - \mu N$ commutes with the number operator, all the correlation functions with a different number of creation and annihilation operators vanish. We adopt the usual statistical mechanics notation by which $a(f)$, the analog of (3.3), is antilinear in f, $a^*(f) = a(f)^*$ and $[a(f), a^*(g)]_\mp = (f, g)$, where $[\,,\,]_\mp$ denotes the commutator/anticommutator and the upper/lower choice refers to the boson/fermion case.

One easily proves that

$$e^{w\,H(\mu)}\, a^*(f)\, e^{-w\,H(\mu)} = e^{-w\mu}\, a^*(e^{w\,h}\, f), \quad w = i\,t,\ -\beta, \tag{13.1}$$

where h is the restriction of H to the one particle subspace. By using the above equation one can easily compute the two point function ($z \equiv e^{\beta\mu}$, $Z = \operatorname{Tr} e^{-\beta H(\mu)}$)

$$Z\,\Omega(a^*(f)\,a(g)) = \operatorname{Tr}\left(e^{-\beta H(\mu)}\,a^*(f)a(g)\right) =$$

$$= z\,\operatorname{Tr}\left(a^*(e^{-\beta h}\, f)\, e^{-\beta\,H(\mu)}a(g)\right) = z\,\operatorname{Tr}\left(e^{-\beta\,H(\mu)}a(g)\,a^*(e^{-\beta h}\, f)\right) =$$

$$= z\,\operatorname{Tr}\left(e^{-\beta\,H(\mu)}\{[a(g),\, a^*(e^{-\beta h}\, f)]_\mp \pm a^*(e^{-\beta h}f)a(g)\}\right) =$$

---

[139] Here we give a short and simplified account; for a mathematically more complete treatment, see O. Bratteli and D.W. Robinson, loc. cit., Vol. II, Sects. 5.2.4, 5.2.5.

F. Strocchi: *Fermi and Bose Gas at Non-zero Temperature*, Lect. Notes Phys. **732**, 151–157 (2008)
DOI 10.1007/978-3-540-73593-9_13

$$= zZ(g, e^{-\beta h} f) \pm z Z \,\Omega(a^*(e^{-\beta h} f)\, a(g)).$$

In conclusion, one has

$$\Omega(a^*((1 \mp z e^{-\beta h})\, f)\, a(g)) = z\,(g, e^{-\beta h} f),$$

i.e., letting $f \to (1 \mp z e^{-\beta h})^{-1} f$,

$$\Omega(a^*(f)\, a(g)) = (g,\, z e^{-\beta h}(1 \mp z e^{-\beta h})^{-1} f). \tag{13.2}$$

By a similar trick, one can easily compute the $2n-$point functions and prove that they can be expressed in terms of products of two point functions. A state with this property is called a *quasi free state*. Furthermore, the correlation function of a product containing a different number of a^* and a vanishes, because $\exp(-\beta H(\mu))$ commutes with the number operator and by computing the trace on a basis of eigenvectors of N, each matrix element vanishes.

Equation (13.2) yields in particular the expectations

$$< n_k > = \Omega(a^*(k)\, a(q)) = \delta_{k,q} \frac{e^{-\beta\,(\omega(k)-\mu)}}{1 \mp e^{-\beta\,(\omega(k)-\mu)}}, \tag{13.3}$$

which are the basis of the elementary treatment of the free Bose and Fermi gas, but (13.2) provides much more detailed information since it determines all the correlation functions.

2. Free Fermi Gas in the Thermodynamical Limit

We start by discussing the thermodynamical limit of the finite volume dynamics

$$\alpha_t^V (a(g)) = a(e^{ith_V} g), \quad g \in L^2(V),$$

where the label V has been spelled out to distinguish quantities in the volume V. Now, since $a(f)^2 = 0$, $\|a(f)\|^4 = \|(a(f)^*a(f))^2\| = \|a(f)^*\{a(f), a(f)^*\}a(f)\| = \|f\|^2\|a(f)\|^2$, i.e. $\|a(f)\| = \|f\|$. Then, since

$$\|\alpha_t^V (a(g)) - a(e^{ith}g)\| = \|(e^{ith_V} - e^{ith})g\| \xrightarrow[V\to\infty]{} 0,$$

the finite volume dynamics converges uniformly to the dynamics defined by

$$\alpha_t(a(g)) = a(e^{ith} g). \tag{13.4}$$

For this result a crucial role is played by the fact that $a(g), a(g)^*$ are bounded operators and that the free Fermi algebra is generated by them through products, linear combinations and norm closures.

We can now discuss the thermodynamical limit of the Gibbs (quasi free) states given by (13.2), with a label V understood.

It is not difficult to see that the correlation functions converge as $V \to \infty$. In particular the limit of the two point function is given by (for simplicity we put the fermion mass $m = 1/2$)

$$\omega(a^*(f)\,a(g)) = (2\pi)^{-s} \int d^s p\, \bar{\tilde{g}}(p)\, \tilde{f}(p)\, z\, e^{-\beta p^2} \left(1 + z e^{-\beta p^2}\right)^{-1}, \qquad (13.5)$$

$\forall f, g \in L^2(\mathbf{R}^s)$. It is also easy to see that in the infinite volume limit one has a quasi free state.

The chemical potential μ, which enters in z, is determined by the condition that the average density

$$\rho(\beta, z) = (4\pi^2\beta)^{-s/2} \int d^s x\, z\, e^{-x^2} \left(1 + z e^{-x^2}\right)^{-1}$$

takes the given value $\bar{\rho}$.[140] In the limit of zero temperature, $(\beta \to \infty)$, one has

$$\rho(\infty, \mu) = (2\pi)^{-s} \int_{p^2 \leq \mu} d^s p$$

and one recovers the analog of (7.8), with $\mu = k_F^2$, i.e. the one-particle states with $p^2 \leq \mu$ are occupied (*Fermi sphere*).

The GNS representation defined by the state (13.5) has a Fock type interpretation in terms of occupation numbers of particles and "holes". For this purpose, one introduces new annihilation and creation operators ($f, g \in L^2(\mathbf{R}^s)$)

$$a_\omega(f) = a(\sqrt{1-T}\,f) \otimes 1 + \theta \otimes a^*(K\sqrt{T}f), \qquad (13.6)$$

$$a_\omega^*(g) = a^*(\sqrt{1-T}\,g) \otimes 1 + \theta \otimes a(K\sqrt{T}g), \qquad (13.7)$$

where T is the positive self-adjoint bounded operator, $||T|| \leq 1$, defined by

$$\omega(a^*(f)\,a(g)) = (T^{1/2}g, T^{1/2}f),$$

θ is an operator which anticommutes with a, a^* and K is an antilinear involution $(Kf, Kg) = (g, f)$. Then, by introducing the state $\Omega_\omega \equiv \Omega_F \otimes \Omega_F$ on the a_ω, a_ω^*, where Ω_F is the Fock vacuum on the a, a^*, and $\theta\Omega_F = \Omega_F$ (this requirement fully determines θ), we have

$$\omega(a^*(f)\,a(g)) = (\Omega_F \otimes \Omega_F)(a_\omega^*(f)\,a_\omega(g)). \qquad (13.8)$$

At zero temperature, T is the multiplication by the characteristic function of the Fermi sphere (in momentum space) and $a_\omega(f)$ has the physical interpretation of destroying a particle outside the Fermi sphere, with wave function $\sqrt{1-T}\,f$ and of creating a "hole" inside the Fermi sphere with wave function $K\sqrt{T}\,f$.

The above equation (13.8) displays the general properties of a KMS state at inverse temperature β, on the Von Neumann algebra $\pi_\omega(\mathcal{A})''$ generated by the a_ω, a_ω^*.

---

[140] For the explicit inversion see A. Leonard, Phys. Rev. **175**, 221 (1968).

The representation is *reducible*; in fact the Von Neumann algebra $\pi_\omega'(\mathcal{A})$ generated by the operators

$$a'_\omega(f) = \mathbf{1} \otimes a(\sqrt{1-T}\,f) + a^*(K\,\sqrt{T}f) \otimes \theta, \qquad (13.9)$$

$$a^{*'}_\omega(g) = \mathbf{1} \otimes a^*(\sqrt{1-T}\,g) + a(K\,\sqrt{T}\,g) \otimes \theta, \qquad (13.10)$$

commutes with the Von Neumann algebra $\pi_\omega(\mathcal{A})''$ generated by a_ω, a^*_ω.

The representation is *primary*. In fact, since $\pi_\omega(\mathcal{A})'' \subseteq \pi_\omega'(\mathcal{A})'$, we have

$$\pi_\omega(\mathcal{A})' \cap \pi_\omega(\mathcal{A})'' \subseteq \pi_\omega(\mathcal{A})' \cap \pi_\omega'(\mathcal{A})' = (\pi_\omega(\mathcal{A})'' \cup \pi_\omega'(\mathcal{A}))'$$

and since $\pi_\omega(\mathcal{A})'' \cup \pi_\omega'(\mathcal{A})$ is a doubled fermionic canonical algebra, which is irreducibly represented by ω, its commutant consists of multiples of the identity.

Furthermore, the two equations

$$\pi_\omega(\mathcal{A})'' \cup \pi_\omega'(\mathcal{A}) = \mathcal{B}(\mathcal{H}),$$

$$\pi_\omega(\mathcal{A})'' \cup \pi_\omega(\mathcal{A})' = (\pi_\omega(\mathcal{A}) \cap \pi_\omega(\mathcal{A})')' = \mathcal{B}(\mathcal{H})$$

imply

$$\pi_\omega(\mathcal{A})' = \pi_\omega'(\mathcal{A}). \qquad (13.11)$$

Since $\pi_\omega'(\mathcal{A})$ is isomorphic to $\pi_\omega(\mathcal{A})''$, the Von Neumann algebra $\pi_\omega(\mathcal{A})''$ is *isomorphic to its commutant*.

It is an instructive exercise to explicitly derive the properties 1-3 listed in Sect. 12.3, for this specific example.

3. Free Bose Gas in the Thermodynamical Limit

In the Bose case the thermodynamical limit is much more delicate and interesting.

First, to discuss the thermodynamical limit of the finite volume dynamics, one must use the Weyl operators and, as already seen in Sect. 7.2, α_t^V does not converge to α_t in the uniform (or norm) topology on the quasi local algebra generated by the local Weyl operators. However, it is not difficult to see that α_t^V converges strongly to α_t on the algebra generated by the L^2-delocalized Weyl operators $U(f)$, $V(g)$, f, g real and $\in L^2(\mathbf{R}^s)$.

More interesting is the thermodynamical limit of the finite volume Gibbs states, defined by (13.2), since it displays the occurrence of a "gas-liquid" phase transition (even if the system is free). For this purpose, we note that in the finite volume V (putting $2m = 1$)

$$\frac{\Omega(N)}{V} = \frac{z}{V(1-z)} + \frac{1}{V}\sum_{k\neq 0}\frac{z}{e^{\beta k^2} - z}, \qquad (13.12)$$

where $z = z(\beta, V)$ has to be chosen in such a way that $\Omega(N)/V = \bar{\rho}$, the pre-assigned fixed density.

Since the first term gives the density of particles at zero momentum, which is therefore a non-negative quantity, one must have

$$0 \leq z(\beta, V) \leq 1.$$

Furthermore, the second term on the r.h.s. of (13.12) is an increasing function of z, which, in the thermodynamical limit, is given by $(z(\beta) \equiv z(\beta, \infty))$

$$\frac{z(\beta)}{(2\pi)^{s/2}} \int d^s k \, \frac{1}{e^{\beta k^2} - z(\beta)} \leq \frac{1}{(2\pi)^{s/2}} \int d^s k \, \frac{1}{e^{\beta k^2} - 1} \equiv \rho'(\beta).$$

Therefore, if the given density $\bar\rho$ is greater than $\rho'(\beta)$, in the thermodynamical limit $z(\beta, V)$ must approach 1 in such a way that

$$\rho_0(\beta, V) \equiv \frac{z(\beta, V)}{V(1 - z(\beta, V))} \xrightarrow{V \to \infty} \bar\rho - \rho'(\beta) \equiv \rho_0(\beta) \neq 0,$$

i.e. as $1 - (\rho_0(\beta) \, V)^{-1}$. For a given density $\bar\rho$, the critical temperature T_c is defined by the equation $\bar\rho = \rho'(\beta)$; it is therefore the temperature at which the given density $\bar\rho$ coincides with the maximum value of the second term on the r.h.s. of (13.12). Thus, since $\rho'(\beta)$ is an increasing function of the temperature, for any $T > T_c$, it is always possible to choose a function $z(\beta, V)$ in such a way that, in the thermodynamical limit, the r.h.s. of (13.12) yields $\bar\rho$, i.e. the equation

$$\bar\rho = z(\beta)(2\pi)^{-s/2} \int d^s k \, (e^{\beta k^2} - z(\beta))^{-1}$$

always has a solution for $z(\beta)$, with $z(\beta) < 1$.

On the other hand, for $T < T_c$ one has $\rho_0(\beta) \neq 0$ and, for large V, $z(\beta, V) \sim 1 - (\rho_0(\beta) \, V)^{-1}$.

In conclusion, in the thermodynamical limit, (13.2) gives (for simplicity we consider the case $s = 3$)

$$\omega(a^*(f) \, a(g)) = (2\pi)^{-3} \int d^3 k \, \bar{\tilde{g}}(k) \, \tilde{f}(k) \, z(\beta) \, (e^{\beta k^2} - z(\beta))^{-1}, \quad \beta < \beta_c,$$

$$\tag{13.13}$$

$$= \rho_0(\beta) \tilde{f}(0) \, \bar{\tilde{g}}(0) + (2\pi)^{-3} \int d^3 k \, \tilde{f}(k) \, \bar{\tilde{g}}(k)(e^{\beta k^2} - 1)^{-1}, \quad \beta > \beta_c. \quad (13.14)$$

Thus, below the critical temperature one has a condensation of particles in the $k = 0$ state (*Bose-Einstein condensation*), i.e. a phase transition between the gas and the "liquid" phase.

This transition is indeed observed for liquid He^4, at the critical temperature of 2.18^o K, not so far from the prediction of the free model discussed above, which gives a critical temperature of 3.14^o K for the density of the liquid Helium (for more information, see e.g. K. Huang, *Statistical mechanics*, Wiley 1987).

4. Bose-Einstein Condensation and Symmetry Breaking

Below the critical temperature the equilibrium state ω defined by the infinite volume limit of (13.2) does not satisfy the cluster property, since $\omega(a(f)) = 0$ and on the other hand, by putting $g_a(x) \equiv g(x + a)$, one has

$$\lim_{|a| \to \infty} \omega(a^*(f)\, a(g)) = \rho_0(\beta)\tilde{f}(0)\,\overline{\tilde{g}}(0),$$

(the second term on the r.h.s. of (13.14) vanishes in the limit by the Riemann-Lebesgue lemma).

Thus, below the critical temperature, ω can be decomposed into primary states, which can be shown to be labeled by an angle θ and to exhibit the *spontaneous breaking of gauge transformations* (defined in Sect. 7.2)

$$\omega_\theta(a(g)) = \sqrt{\rho_0}\, e^{i\theta}\, \overline{\tilde{g}}(0). \tag{13.15}$$

Such a decomposition can be obtained by appealing to general methods[141]. It can also be obtained in a rather elementary way by introducing a symmetry breaking coupling with a constant external field $j_{ext} = j\, e^{i\theta}$, $j > 0$, according to the Bogoliubov strategy[142] discussed in Chap. 10.

The corresponding finite volume Hamiltonian H (the suffix V is omitted for simplicity) then is given by

$$H = \sum_k (k^2 - \mu)\, a^*(k)\, a(k) + j(a_0^*\, e^{i\theta} + a_0\, e^{-i\theta})\sqrt{V},$$

where $a_0 \equiv a(k = 0)$, and can be easily brought to diagonal form

$$H = \sum_k (k^2 - \mu)A^*(k)\, A(k) + j^2 V/\mu,$$

in terms of the following new annihilation and creation operators

$$A(k) = a(k),\ \text{for}\ k \neq 0; \quad A(k = 0) = a_0 - (j/\mu)\, e^{i\theta}\, \sqrt{V}.$$

By proceeding as in Sect. 13.1, one easily gets, for the equilibrium state ω_j, $\omega_j(A(f)) = 0$, which implies

$$\omega_j(a(f)) = \overline{\tilde{f}}(0)\, \omega_j(a_0/\sqrt{V}) = \overline{\tilde{f}}(0)\, e^{i\theta}\, j/\mu. \tag{13.16}$$

Furthermore, the analog of (13.12) gives

$$\rho \equiv V^{-1}\omega_j\Big(\sum_k a^*(k)\, a(k)\Big) = V^{-1}\omega_j\Big(\sum_k A^*(k)\, A(k)\Big) + j^2/\mu^2$$

[141] See J. Cannon, Comm. Math. Phys. **29**, 89 (1973) and O. Bratteli and D.W. Robinson, loc. cit., Vol. II, pp. 72-73.

[142] N.N. Bogoliubov, *Lectures on Quantum Statistics*, Vol. 2, Part 1, Gordon and Breach 1970.

$$= \frac{z}{V(1-z)} + \frac{1}{V} \sum_{k \neq 0} \frac{z}{e^{\beta k^2} - z} + \frac{\beta^2 j^2}{(\ln z)^2}. \qquad (13.17)$$

The fugacity $z = z(\beta, V, j)$ has to be chosen in such a way that in the limit $j \to 0$, taken after the thermodynamical limit, one gets $\rho = \bar{\rho}$, the preassigned density.

Now, the third term in the r.h.s. of (13.17) is an increasing function of z, $0 \leq z \leq 1$, which vanishes when $z \to 0$ and tends to infinity when $z \to 1$. Thus, for any given density $\bar{\rho}$, the equation

$$\frac{1}{(2\pi)^{3/2}} \int d^3k \, (e^{\beta k^2} - 1) + \frac{\beta^2 j^2}{(\ln z)^2} = \bar{\rho} \qquad (13.18)$$

always has a solution $z = z(\beta, \infty, j) < 1$, and consequently the first term on the r.h.s. of (13.17) vanishes in the thermodynamical limit. Then, in such a limit, putting $\rho_0(\beta, j) \equiv \beta^2 j^2/(\ln z)^2$, one gets

$$\omega_j(a(f)) = (\rho_0(\beta, j))^{1/2} \, e^{i\theta} \, \overline{\tilde{f}(0)}. \qquad (13.19)$$

Now we discuss the limit $j \to 0$ by distinguishing two cases:

1) $T > T_c$. In this case, for any given $\bar{\rho}$, (13.18), with $j = 0$, always has a solution for $z = z(\beta)$, with $0 < z(\beta) < 1$. Thus, by choosing z so that $z(\beta, \infty, j) \to z(\beta)$, when $j \to 0$, one gets in this limit

$$\rho(\beta, j) \to \bar{\rho}, \quad \rho_0(\beta, j) \to 0.$$

Thus, $\omega_j(a(f)) \to 0$, independently of the way $je^{i\theta} \to 0$, and therefore there is a unique phase.

2) $T < T_c$. In this case, $\bar{\rho} - \rho'(\beta) > 0$ and therefore one must have

$$\rho_0(\beta, j) \equiv \beta^2 j^2/(\ln z)^2 \xrightarrow{j \to 0} \bar{\rho} - \rho'(\beta) \equiv \rho_0(\beta) > 0.$$

This requires to choose z in such a way that $z(\beta, \infty, j) \to 1$, as $j \to 0$; it suffices to take $z(\beta, \infty, j) = \exp\left[-\beta j \, (\rho_0(\beta))^{-1/2}\right]$. Then

$$\lim_{j \to 0} \omega_j(a(f)) = (\rho_0(\beta))^{1/2} \, e^{i\theta} \, \overline{\tilde{f}(0)} \equiv \omega_\theta(a(f)). \qquad (13.20)$$

Thus the limit depends on the phase θ of the external field and one has a one-parameter family of equilibrium states ω_θ, $\theta \in [0, 2\pi)$. As it is easy to see, all such states satisfy the cluster property; therefore, they are primary states on the Weyl algebra and define pure phases. Clearly, each state ω_θ is not invariant under gauge transformations, which are therefore spontaneously broken in each representation defined by ω_θ.

14 Quantum Fields at Non-zero Temperature

The general structure discussed above provides a neat and unique prescription for the quantization of relativistic fields at non-zero temperature (*thermofield theory*).

For simplicity we consider the case of a relativistic scalar field (see Example 2.1)

$$\phi(x) = (2\pi)^{-3/2} \int d^3k \, (2\omega_k)^{-1} \, [a_k \, e^{-ikx} + a_k^* e^{ikx}], \qquad (14.1)$$

where $kx = k_0 x_0 - \mathbf{k} \cdot \mathbf{x}$, $k_0 = \omega_k = (\mathbf{k}^2 + m^2)^{1/2}$ and the annihilation and creation operators have been so normalized that they obey the canonical commutation relations (CCR) in relativistically covariant form

$$[a_k, a_{k'}^*] = 2\omega_k \, \delta(\mathbf{k} - \mathbf{k}'), \qquad [a_k, a_{k'}] = 0. \qquad (14.2)$$

This fixes the algebraic structure.

The equilibrium (gauge invariant) state at inverse temperature β is characterized by the KMS condition, (12.4),

$$< a_k^* \, a_q >_\beta \equiv \omega_\beta(a_k^* \, a_q) = \omega_\beta(a_q \, \alpha_{-i\beta}(a_k^*)) = e^{-\beta \omega_k} < a_q \, a_k^* >_\beta,$$

where for simplicity we have considered the case of zero chemical potential.

On the other hand, the CCR give

$$< a_k^* \, a_q >_\beta = < a_q \, a_k^* >_\beta - 2\omega_k \delta(\mathbf{k} - \mathbf{q}).$$

In conclusion, one has

$$< a_k^* \, a_q > = 2\omega_k \, N(\omega_k) \, \delta(\mathbf{k} - \mathbf{q}), \qquad (14.3)$$

$$N(\omega_k) \equiv e^{-\beta \omega_k} \, (1 - e^{-\beta \omega_k})^{-1} = (e^{\beta \omega_k} - 1)^{-1}.$$

Then, the CCR and the above equation give for the two point function of ϕ, putting $z \equiv x - y$,

$$< \phi(x) \, \phi(y) >_\beta = \frac{1}{(2\pi)^3} \int \frac{d^3k}{(2\omega_k)} \, e^{-i\mathbf{k} \cdot \mathbf{z}} \, [e^{i\omega_k z_0} N(\omega_k) + e^{-i\omega_k z_0} (1 + N(\omega_k))].$$

F. Strocchi: *Quantum Fields at Non-zero Temperature*, Lect. Notes Phys. **732**, 159–160 (2008)
DOI 10.1007/978-3-540-73593-9_14

By using the identity
$$N(\omega) = -(1 + N(-\omega)),$$
one can cast the above two point function in the form[143]

$$< \phi(x)\,\phi(y) >_\beta = (2\pi)^{-4} \int d^4k\, \delta(k^2 - m^2) e^{-ikz}\, \varepsilon(k_0)\,(1 + N(k_0)). \quad (14.4)$$

The relativistic spectral condition yields a larger analyticity domain than in the non-relativistic case; in fact, the two point function has an analytic continuation to the domain $\{z \in \mathbf{C}^4; \mathrm{Im}z \in V_+ \cap (\beta, \mathbf{0}) + V_-\}$ where $V_\pm$ denote the forward and backward cones (*relativistic KMS condition*)[144].

Historically, the quantization of fields at non-zero temperature has been done with different strategies, based on the functional integral approach. The same results can be obtained more directly by exploiting the KMS condition. For example, the so-called *imaginary time* (Matsubara) formulation can be obtained if one i) analytically continues the correlation functions to purely imaginary time and ii) introduces a complex time ordering of products of operators with respect to a fixed complex time contour.

As an example we consider the two point function $< A\,A_z > \equiv \Omega(A\,A_z)$, where A denotes a field variable at time zero and A_z, $z = i\tau$, $-\beta < \tau < \beta$, the corresponding variable after an imaginary time translation (as in (12.5)). A (complex) time ordered expectation is defined by

$$\Delta^T(\tau) = \theta(\tau)\,\mathrm{Tr}\,(e^{-\beta H}\,A\,A_z) + \theta(-\tau)\,\mathrm{Tr}(e^{-\beta H}\,A_z\,A), \quad (14.5)$$

where θ denotes the Heaviside step function and $-\beta < \mathrm{Im}z = \tau < \beta$. Thus, the KMS condition (12.4) gives

$$\Delta^T(\tau + \beta) = \Delta^T(\tau).$$

The periodicity implies that only discrete frequencies occur in the Fourier transform of Δ^T.

---

[143] J. Bros and D. Buchholz, Z. Phys. C-Particles and Fields **55**, 509 (1992); Nucl. Phys. **B429**, 291 (1994).

[144] J. Bros and D. Buchholz, loc. cit., previous footnote.

15 Breaking of Continuous Symmetries. Goldstone's Theorem

For a long time, the mechanism of spontaneous breaking of continuous symmetries has been recognized to be at the basis of many collective phenomena and in particular of phase transitions in statistical mechanics; recently, it has played a crucial role in the developments of theoretical physics, both at the level of many body physics and for the unification of elementary particle interactions.

For relativistic systems and more generally for systems with short range dynamics, the clarification of the mechanism has been achieved to a high level of rigor and formalized in the so-called *Goldstone's theorem*[145]. The result is that the conditions for the applicability of the conclusions, a subject of discussions in the early developments, are now out of question.

The important point is that the Goldstone theorem provides non-perturbative exact information on the excitation spectrum, since it predicts the low momentum behaviour of the energy, $\omega(\mathbf{k}) \to 0$, as $\mathbf{k} \to 0$, of the elementary excitations (*Goldstone bosons*) associated with the broken symmetry generators. The examples are many and they appear in different branches of physics, like the spin waves in the theory of ferromagnetism, the Landau phonons in the theory of superfluidity, the phonon excitations in crystals, the pions as Goldstone particles of chiral symmetry breaking etc.

In this chapter, we first give the simple "heuristic proof" of the Goldstone theorem (in the zero temperature case) without caring about subtle mathematical points; the aim is to show in a simple way the connection between symmetry breaking of continuous symmetry and absence of energy gap.

The idea is that if the ground state ω of an extended system is not symmetric under a continuous symmetry $\beta^\lambda, \lambda \in \mathbf{R}$, leaving the Hamiltonian

---

[145] J. Goldstone, A. Salam and S. Weinberg, Phys. Rev. **127**, 965 (1962); D. Kastler, D.W. Robinson and J.A. Swieca, Comm. Math. Phys. **2**, 108 (1966); D. Kastler, Broken Symmetries and the Goldstone Theorem in Axiomatic Field theory, in *Proceedings of the 1967 International Conference on Particles and Fields*, C.R. Hagen et al. eds., Interscience 1967; J.A. Swieca, Goldstone theorem and related topics, in *Cargése Lectures in Physics*, Vol. 4, D. Kastler ed., Gordon and Breach 1970; R.F. Streater, Spontaneously broken symmetries, in *Many degrees of freedom in Field Theory*, L. Streit ed., Plenum Press 1978.

F. Strocchi: *Breaking of Continuous Symmetries. Goldstone's Theorem*, Lect. Notes Phys. **732**, 161–176 (2008)
DOI 10.1007/978-3-540-73593-9_15

invariant, then the states ω_{β_R}, obtained from ω by applying a symmetry transformation β_R localized in a region of radius R, have the same energy of the ground state except for boundary terms. Since the symmetry is continuous one can smooth the transition region so that the boundary terms vanish when $R \to \infty$, and so does the energy of the states ω_{β_R}.

To formalize the idea, one abstracts from the Lagrangean (or Hamiltonian) formulation the information (Noether's theorem) that the invariance under a continuous symmetry β^λ implies the existence of a conserved current, whose charge density generates the symmetry transformation: namely, $\forall A \in \mathcal{A}_L$(=the local algebra), the infinitesimal variation under β^λ is given by[146]

$$\delta A = d\beta^\lambda(A)/d\lambda \,|_{\lambda=0} = i \lim_{R \to \infty} [Q_R, A], \tag{15.1}$$

$$Q_R = \int_{|\mathbf{x}| \leq R} d^s x \, j_0(\mathbf{x}, 0) \tag{15.2}$$

$$\partial_t j_0(\mathbf{x}, t) + \operatorname{div} \mathbf{j}(\mathbf{x}, t) = 0. \tag{15.3}$$

A relevant point is that the above infinitesimal generation of the symmetry holds for a subalgebra $\mathcal{A}_0$ of $\mathcal{A}$, containing $\mathcal{A}_L$, stable under time translations. The above equations encode the essential features of a continuous symmetry without relying on the definition of the Lagrangean.

For symmetries which commute with space and time translations, the current transforms covariantly under space and time translations

$$U(\mathbf{a}, \tau) j_\mu(\mathbf{x}, t) U(\mathbf{a}, \tau)^{-1} = j_\mu(\mathbf{x} + \mathbf{a}, t + \tau), \quad \mu = 0, 1, \ldots \tag{15.4}$$

Briefly, an internal continuous symmetry β^λ satisfying the above properties is said to be *locally generated by a charge density* associated with a conserved current.

15.1 The Goldstone's Theorem

The (heuristic) version of the Goldstone theorem[147], which does not use manifest relativistic covariance, says ($A = A^*$ covers the general case since any B is $= B_1 + iB_2$, $B_i = B_i^*$):

Theorem 15.1. *(Goldstone) If*

I. $\beta^\lambda, \lambda \in \mathbf{R}$ *is a one-parameter internal symmetry group, i.e.*

$$[\beta^\lambda, \alpha_{\mathbf{x}}] = 0, \quad [\beta^\lambda, \alpha_t] = 0, \quad \forall \lambda \in \mathbf{R}, \, \mathbf{x} \in \mathbf{R}^s, \, t \in \mathbf{R} \tag{15.5}$$

---

[146] Equation (15.1) may be understood to hold as a bilinear form on a dense set of states, in each relevant representation; actually, all what is needed is its validity on the ground state.

[147] The non-relativistic version has been discussed in particular by R.V. Lange, Phys. Rev. Lett. **14**, 3 (1965); Phys. Rev. **146**, 301 (1966) and by J.A. Swieca, Comm. Math. Phys. **4**, 1 (1967).

II. β^λ is locally generated by a charge in the sense of (15.1-4) on a subalgebra $\mathcal{A}_0$ of $\mathcal{A}$, stable under time evolution

III. β^λ is spontaneously broken in a representation π defined by a translationally invariant ground state Ψ_0, i.e. there exists a (self-adjoint) $A \in \mathcal{A}_0$ such that

$$< \delta A >_0 = i \lim_{R \to \infty} < [Q_R, A] >_0 = b \neq 0, \qquad (15.6)$$

then, in the subspace generated by the vectors $Q_R \Psi_0$, $R \in \mathbf{R}$, **the energy spectrum at zero momentum cannot have a gap** (with respect to the ground state energy).

Proof.   Information on the energy momentum spectrum of the state $Q_R \Psi_0$ is provided by the support of the Fourier transform of the matrix elements $(A\Psi_0, U(\mathbf{x}) U(t) Q_R \Psi_0)$, or, in particular, of their imaginary part. This follows from the spectral theorem for $U(\mathbf{x}) U(t)$ (or by inserting a complete set of improper eigenstates of energy and momentum, see footnote below).

Thus, we are led to analyze the Fourier transform of

$$J(\mathbf{x}, t) \equiv i < [j_0(\mathbf{x}, t), A] >_0 = 2 \operatorname{Im} < A j_0(\mathbf{x}, t) >_0 . \qquad (15.7)$$

By using the property that β^λ commutes with α_t and that it is generated by Q_R on an algebra stable under time translations, we have $(Q_R(t) = U(t) Q_R U(t)^{-1})$,

$$i \lim_{R \to \infty} < [Q_R(t), A] >_0 = i \lim_{R \to \infty} < [Q_R, \alpha_{-t}(A)] >_0 = < \delta(\alpha_{-t}(A)) >_0$$

$$= < \alpha_{-t}(\delta A) >_0 = < \delta A >_0 = i \lim_{R \to \infty} < [Q_R, A] >_0 = b. \qquad (15.8)$$

Then, we have

$$\lim_{R \to \infty} \int_{|\mathbf{x}| \leq R} d^s x \, J(\mathbf{x}, t) = b,$$

namely, by Fourier transforming in $\mathbf{x}$ and t,

$$\lim_{\mathbf{k} \to 0} \tilde{J}(\mathbf{k}, \omega) = (2\pi)^{-1} b \, \delta(\omega). \qquad (15.9)$$

This is incompatible with an energy gap at $\mathbf{k} \to 0$.[148]

The above standard (heuristic) argument would completely settle the statement of Goldstone's theorem (apart from somewhat pedantic mathematical polishing) were it not for the existence of physically interesting models which seem to evade the conclusions of the theorem. The attention on these

---

[148] In fact, if $|\mathbf{k}, \omega(\mathbf{k})_l, l >$ denote the improper eigenstates of momentum and energy, with l the additional quantum numbers needed to remove possible degeneracies, then $\lim_{\mathbf{k} \to 0} \omega_l(\mathbf{k}) \geq \mu > 0$ implies that

$$\lim_{\mathbf{k} \to 0} \tilde{J}(\mathbf{k}, \omega_l) = \lim_{\mathbf{k} \to 0} 4\pi^2 \, 2 \operatorname{Im} \sum_l < A\Psi_0 \, | \, \mathbf{k}, \omega_l, l >< \mathbf{k}, \omega_l, l \, | \, j_0(0,0)\Psi_0 >$$

cannot satisfy (15.9).

examples arose especially in the early sixties in connection with attempts to interpret the $SU(3)$ eightfold way as a spontaneously broken symmetry, notwithstanding the absence of the corresponding Goldstone bosons. Among such examples we mention the BCS model of superconductivity, where the $U(1)$ internal symmetry is spontaneously broken with energy gap, the breaking of the Galilei symmetry in Coulomb systems, which is accompanied by the plasmon energy gap, the Higgs mechanism in the Coulomb gauge and the breaking of the axial $U(1)$ symmetry in quantum chromodynamics (QCD) (the so-called $U(1)$ problem), both with no corresponding massless Goldstone bosons, etc.

Clearly, in such examples some of the assumptions of the theorem must fail, but the long discussions on the possible mechanisms for evading the conclusions of the theorem seem to have led more to a series of catchwords or perturbative prescriptions, rather than to a sharp and clear identification of the crucial points. For non-relativistic systems, the standard explanation for the presence of an energy gap is that the Coulomb potential leads to a shift of energy (at $\mathbf{k} \to 0$), by a mechanism advocated on the basis of clever ad hoc approximations, rather than in terms of a general non-perturbative mechanism. The problem with such an explanation is that long range correlations and interactions, which always occur when there are massless particles, do not invalidate the applicability of the theorem in relativistic local quantum field theory. The standard explanation of the Higgs mechanism relies on the perturbative expansion, and for the $U(1)$ problem the standard explanation, in terms of the chiral anomaly and instanton calculations, does not seem to provide a general clearcut solution and some questions remain open.[149]

The above considerations justify a critical analysis of the hypotheses of the theorem and their verification. As we shall see, the standard explanations of the "evasion" of the theorem are somewhat incomplete, if not misleading, since they seem to overlook the basic delicate points and miss the general mechanism.

15.2 A Critical Look at the Hypotheses of Goldstone Theorem

The importance and usefulness of the Goldstone theorem is mainly that of providing non-perturbative information on the energy spectrum of an infinite system. For this purpose, it is crucial to be able to verify its assumptions without having to solve the full dynamical problem. We shall therefore critically discuss the hypotheses of the theorem and their possible verification; as a result, we shall discover the general mechanism which is at the basis of the phenomenon of spontaneous breaking of a continuous symmetry accompanied by an energy gap, in all the examples mentioned above.

---

[149] For a critical discussion, see F. Strocchi, *Selected Topics on the General Properties of Quantum Field theory*, World Scientific (1993), Sect. 7.4 iv.

I. Symmetry of the Dynamics

At a formal level, the existence of an internal symmetry is inferred from the invariance of the formal Hamiltonian (or Lagrangean) which (formally) defines the model.

Now, the commutation of the symmetry β^λ with the space translations $\alpha_{\mathbf{x}}$ is a kinematical property which is easily checked, once the action of β^λ on the canonical variables (or on the observables) is specified.

Less obvious is the check of the commutation of β^λ with the time translations α_t, since in general the infinite volume dynamics is not explicitly known.

Proposition 15.2. *If the finite volume dynamics α_t^V, defined by the finite volume Hamiltonian H_V, converges to the infinite volume dynamics α_t in the norm topology, then*

$$\beta^\lambda \alpha_t^V = \alpha_t^V \beta^\lambda \tag{15.10}$$

implies

$$\beta^\lambda \alpha_t = \alpha_t \beta^\lambda. \tag{15.11}$$

Proof. In fact, *- automorphisms of a C^*-algebra are norm preserving and therefore continuous in the norm topology

$$\beta^\lambda \alpha_t(A) = \beta^\lambda(\alpha_t - \alpha_t^V)(A) + \alpha_t^V \beta^\lambda(A) \xrightarrow[V \to \infty]{} \alpha_t \beta^\lambda(A).$$

Thus, the check of (15.11) is reduced to the invariance of the finite volume Hamiltonian and the current wisdom is essentially correct.

II. Generation of the Symmetry by a Local Charge

Much more problematic and subtle is condition II, the precise formulation of which involves properties with important physical consequences.

i) *Local charge as an integral of a density*
First, for technical reasons (see below), it is convenient to smooth out the sharp boundary in (15.2), by introducing a C^∞ function of compact support[150] (for simplicity we omit the boldface notation for the variable $x \in \mathbf{R}^s$)

$$f_R(x) = f(|x|/R), \quad f \in \mathcal{D}(\mathbf{R}), \tag{15.12}$$

$$f(x) = 1, \ \text{for } |x| \leq 1, \ \ f(x) = 0, \ \text{for } |x| \geq 1 + \varepsilon,$$

and replace the definition of $Q_R(t)$ in (15.2), (15.6), by

$$Q_R(t) = \int dx \, f_R(x) \, j_0(x, t) \equiv j_0(f_R, t). \tag{15.13}$$

---

[150] As we shall discuss below, for relativistic systems also a smearing in time is necessary to cope with the ultraviolet singularities.

Even with such a *proviso*, the limit $R \to \infty$, i.e. the formal integral $Q(t) = \int dx\, j_0(x, t)$, does not exist and therefore it does not define an operator, since by (15.4) the current density does not "decrease" for $|x| \to \infty$.

Much better are the convergence properties of the integral of the commutator

$$\mathcal{J}(x, t) = i\, [j_0(x, t),\, A],$$

with a local operator A, since $\mathcal{J}(x, t)$ at least vanishes for $|x| \to \infty$, by asymptotic abelianess and actually has compact support if j_0 and A satisfy the relativistic locality property (Chap. 4, (4.2)).

It is implicit in (15.1) that $\mathcal{J}(x, t)$ must be at least integrable in x. For a mathematical control of the proof, one actually needs that $\mathcal{J}(x, t)$ is absolutely integrable in x for large $|x|$.[151] Thus, one must supplement the condition of local generation by a charge with the **integrability condition of the charge density commutators.** also briefly called *charge integrability condition*.

It means that the ground state expectation values of the charge density commutators are absolutely integrable in x for large $|x|$, as tempered distributions in t, i.e. $\forall g \in \mathcal{S}(\mathbf{R})$

$$< \int dt\, g(t)\, [j_0(x, t),\, A] >_0 \tag{15.14}$$

is absolutely integrable in x for large $|x|$.

Such an integrability condition is satisfied if j_0 and A satisfy the relativistic locality condition, since the smearing with $g(t) \in \mathcal{S}$ can at worse change the compact support in x of $J(x, t)$ to a fast decrease.[152]

More generally, the condition is satisfied by *systems with short range dynamics*, namely if $\forall A, B \in \mathcal{A}_L$, as distributions in t,

$$\lim_{|x| \to \infty} |x|^{s+\varepsilon} < [A_x,\, \alpha_t(B)] >_0 = 0, \tag{15.15}$$

where $s = $ space dimensions, $\varepsilon > 0$. This is the case of spin systems with short range interactions (see Sect. 7.3).

It is worthwhile to note that the charge integrability condition, (15.14), is much weaker than (15.15), since it involves a special operator j_0; as we shall see below, (15.15) fails in models with long range interactions, whereas there are indications that the charge integrability condition holds.

In conclusion, the charge integrability should be taken as part of the definition that β^λ is locally generated by a charge density; the physical meaning

---

[151] G. Morchio and F. Strocchi, Comm. Math. Phys. **99**, 153 (1985); J. Math. Phys. **28** 622 (1987). The crucial role of such a condition for the non-relativistic version of the Goldstone theorem and the need of a careful handling of the distributional and measure theoretical problems do not seem to have been noted in the vast previous literature.

[152] This can be seen, e.g. by using the Jost-Lehmann-Dyson representation.

is that β^λ can be reasonably well approximated by *-automorphisms β_R^λ with good localization properties. As we shall see below, such condition leads to the existence of quasi particles with infinite lifetime in the limit of zero momentum (*Goldstone quasi particles*).

ii) *Local generation by a charge and time evolution*
The really delicate issue (not sufficiently emphasized in the literature), which crucially enters in the proof of the theorem, is the condition that the local generation by a charge, (15.6), holds on an algebra stable under time evolution. An equivalent condition is that Q_R and $Q_R(t) = \alpha_t(Q_R)$ generate the same automorphism.

Equation (15.6) can be easily checked on the time zero algebra generated by the canonical variables at time $t = 0$, since this is a purely kinematical question which involves the CCR or the ACR. The problem is whether (15.6) remains true when A is replaced by $\alpha_t(A)$. This is not trivial to check, because the infinite volume dynamics α_t is not explicitly known and the limit $R \to \infty$ involved in (15.6) may not commute with the infinite volume limit of α_t^V.

The heuristic argument that "since the Hamiltonian commutes with β^λ the charge which generates β^λ is independent of time, i.e. $Q(t) = Q(0)$," and therefore

$$\lim_{R\to\infty} [Q_R(t), A] = \lim_{R\to\infty} [Q_R(0), A]$$

is not correct, because it overlooks the following important points. A global charge as algebraic generator of β^λ does not exist if the symmetry is broken (we have already remarked that the formal integral of j_0 does not define an operator), and one can only speak of a local generation of β^λ in terms of local charges, so that a limit $R \to \infty$ is unavoidably involved in (15.6). The commutation of β^λ with α_t does not imply that the limit of the commutator $[Q_R(t), A]$ is time independent; in fact, the symmetry of the finite volume Hamiltonian

$$\lim_{R\to\infty} [Q_R, H_V] = 0$$

implies the time independence of the limit of the commutator $[Q_R(t), A]$, provided the two limits $R \to \infty$, $V \to \infty$ commute. In fact, in this case one has

$$\lim_{R\to\infty} [\dot{Q}_R(0), A] = i \lim_{R\to\infty} \lim_{V\to\infty} [[H_V, Q_R(0)], A] =$$

$$= i \lim_{V\to\infty} \lim_{R\to\infty} [[H_V, Q_R(0)], A] = 0.$$

These remarks may look pedantic and with little physical relevance, but they actually identify the crucial point which invalidates the heuristic argument and is at the basis of the apparent evasion of the Goldstone theorem by the physically relevant examples mentioned above. As a matter of fact, the interchangeability of the two limits depends on the localization properties of the dynamics, which in turn are governed by the range of the potential. Indeed, for short range interactions the dynamics essentially preserves the localization of the operators, so that the limit $R \to \infty$ is essentially reached for finite R and the interchange of the limits is allowed.

The role of the delocalization effects of the time evolution can be explicitly displayed by working out the implications of the current conservation on the time dependence of the charge commutator

$$[\dot{Q}_R(t),\, A] = -\int dx\, f_R(x)[\operatorname{div} \mathbf{j}(x,t),\, A] = \int dx\, \nabla f_R(x)\, [\mathbf{j}(x,t),\, A]. \tag{15.16}$$

Now, since $\operatorname{supp} \nabla f_R(x) \subset \{R \le |x| \le R(1+\varepsilon)\}$ the time independence of the charge commutator in the limit $R \to \infty$ is governed by the fall off of the commutator $[\mathbf{j}(x,t),\, A]$ for $|x| \to \infty$.

As pointed out by Swieca,[153] the time independence of the charge commutators holds if the time evolution is sufficiently local, i.e. if in s space dimensions $\forall A,\, B \in \mathcal{A}_L$

$$\lim_{|x|\to\infty} |x|^{s-1}[A_x,\, \alpha_t(B)] = 0, \tag{15.17}$$

(*Swieca condition*). In fact, if (15.17) holds, $\forall \delta > 0, \exists L$ such that for $|x| > L$, t in a compact set,

$$|x|^{s-1}|<[\mathbf{j}(x,t),\, A]>_0| < \delta$$

and therefore the r.h.s. of (15.16) is bounded by $(y \equiv x/R)$

$$\delta \int dx\, |\nabla f_R(x)|\, |x|^{1-s} = \delta \int dy\, |\nabla f(y)|\, |y|^{1-s} = \delta\, C.$$

This implies that

$$|<[Q_R(t) - Q_R(0),\, A]>_0| \le \int_0^t dt'\, |<[\dot{Q}_R(t'),\, A]>_0| \le \delta\, C\, t,$$

i.e.

$$\lim_{R\to\infty} <[Q_R(t),\, A]>_0 = \lim_{R\to\infty} <[Q_R(0),\, A]>_0. \tag{15.18}$$

The Swieca condition is clearly satisfied if the dynamics is strictly local, i.e. it maps $\mathcal{A}_L$ into $\mathcal{A}_L$. This is the case if j_μ and A satisfy the relativistic locality condition (see Sect. 4.1). More generally, it is enough that the delocalization induced by the dynamics falls off exponentially, namely, $\forall A,\, B \in \mathcal{A}_L$,

$$\lim_{|\mathbf{x}|\to\infty} |x|^n [A_x,\, \alpha_t(B)] = 0, \quad \forall n \in \mathbf{N}. \tag{15.19}$$

As we have seen, this is the case of free non-relativistic systems (see Sect. 7.2) and the case of spin systems on a lattice with short range interactions (see Sect. 7.3). There are arguments (see below) indicating that property (15.19)

---

[153] J.A. Swieca, Comm. Math. Phys. **4**, 1 (1967).

should hold also for non-relativistic systems with exponentially decreasing interaction potentials.[154]

The above discussion indicates that for systems with sufficiently local dynamics the verification of conditions I, II of the Goldstone theorem is essentially reduced to the symmetry of the finite volume Hamiltonian and to the check that the symmetry is generated by a local charge at equal times. Thus, in these cases the heuristic criteria for the application of the Goldstone theorem are essentially correct.

It is worthwhile to remark that the local properties of the current and of the operator A, which gives rise to the symmetry breaking order parameter, are both crucial for the time independence of (15.6) and may be problematic even in the case of relativistic systems. As a matter of fact, the axial $U(1)$ symmetry of QCD Lagrangean is not generated by a local conserved current in positive gauges and similarly the order parameter, which breaks the gauge symmetry in the Higgs effect in positive gauges, is not given by a local operator.[155]

The discriminating point is not the existence of long range correlations, which may also be present in strictly local theories, but the delocalization induced by the time evolution, typically as a consequence of the non-relativistic approximation or of the gauge fixing condition.

For systems with long range dynamics, the symmetry of the (finite volume) Hamiltonian and the local generation of the symmetry by a local charge at equal times (the standard heuristic criteria for the applicability of the theorem) are not enough to conclude that the hypotheses of the Goldstone theorem are satisfied. This is the way the conclusions of the Goldstone theorem are evaded by the physically relevant examples mentioned above exhibiting a spontaneous symmetry breaking, which satisfies the heuristic criteria but is accompanied by an energy gap. The somewhat mysterious statement that the long range Coulomb potential leads to an energy shift should be interpreted, in the light of the above discussion, as the time dependence of the charge commutators due to the long range delocalization induced by time evolution.

A more explicit discussion of the effect the delocalization induced by long range dynamics, as in the case of Coulomb systems, shall be done below.

iii) *Dynamical delocalization and range of the interaction*
The crucial role of the localization properties of the dynamics for the check of the hypotheses of Goldstone's theorem suggests to get some concrete idea on the relation with the range of the interaction.

---

[154] J.A. Swieca, loc. cit. (1967).

[155] G. Morchio and F. Strocchi, Infrared problem, Higgs phenomenon and long range interactions, in *Fundamental Problems of Gauge Field Theory*, Erice School 1985, G. Velo and A.S. Wightman eds., Plenum Press 1986; F. Strocchi, *Selected Topics on the General Properties of Quantum Field Theory*, World Scientific 1993, Sect. 7.4; G. Morchio and F. Strocchi, J. Phys. A: Math. Theor. **40**, 3173 (2007).

For this purpose, we consider a non-relativistic many body system described by the following finite volume Hamiltonian

$$H_V = H_{0,V} + gH_{int,V} = (1/2m) \int_V dx\, |\nabla\psi(x)|^2 +$$

$$+(g/2) \int_V dx\, dy\, \mathcal{V}(x-y)\, \psi^*(x)\psi^*(y)\psi(y)\psi(x), \qquad (15.20)$$

where $\mathcal{V}(x) = \mathcal{V}(-x)$ denotes a two body interaction potential. To avoid the discussion of short distance singularities, we assume that the potential vanishes in a neighborhood of the origin.

An interaction Hamiltonian of this type, with $\mathcal{V}$ the Coulomb potential, occurs in the theory of non-relativistic Coulomb systems as well as in the time evolution of charged fields in positive gauges, like the Coulomb gauge in quantum electrodynamics.[156]

It is worthwhile to remark that in the case of short range potential the above formal Hamiltonian is supposed to define the dynamics through its finite volume restriction and a suitable limit of the corresponding finite volume dynamics. For long range potentials, like e.g. the Coulomb potential, a counter term has to be added in order to be able to remove the volume cutoff in the equations of motion (on a class of states with enough regularity at space infinity) (*infrared renormalization*).

A convenient possibility[157] is to use the following infrared cutoff Hamiltonian with an infrared counter term

$$H_L = H_0 + gH_{int,L} = (1/2m) \int dx\, |\nabla\psi(x)|^2 +$$

$$+(g/2) \int dx\, dy\, \mathcal{V}_L(x-y)\, \psi^*(x)\, (\,\psi^*(y)\psi(y) - 2\rho_L\,)\, \psi(x), \qquad (15.21)$$

where

$$\mathcal{V}_L(x) \equiv \mathcal{V}(x)\, f_L(x), \quad \rho_L \equiv L^{-3} \int_{|x|<L} dx\, \psi^*(x)\psi(x),$$

f_L is defined in (15.12) and ρ_L has the meaning of an average density, which converges to an element ρ_∞ of the center, in the limit in which the infrared cutoff L is removed, $L \to \infty$, (on a class of states regular at infinity, as explained below). Apart from a (infrared divergent) c-number term which does not affect the commutators, the interaction term can be written as

$$(g/2) \int dx\, dy\, \mathcal{V}(x-y)\, \rho(x)\, \rho(y), \quad \rho(x) \equiv (\psi^*\, \psi)(x) - \rho_L. \qquad (15.22)$$

The corresponding equations of motion are (putting for simplicity $2m = 1$)

$$i\frac{d}{dt}\psi = (-\Delta + g(\mathcal{V}_L * \rho)(x))\psi(x) + O(L^{-3}). \qquad (15.23)$$

---

[156] See e.g J.D. Bjorken and S.D. Drell, *Relativistic Quantum Fields*, McGraw-Hill Book Company 1965, Sect. 15.2.

[157] G. Morchio and F. Strocchi, Ann. Phys. **170**, 310 (1986), esp. Sect. 3.

The effect of the infrared counter term is to subtract the interaction with the average density. The removal of the infrared cutoff requires that, for Coulomb systems in three space dimensions, the density $(\psi^* \psi)(x)$ approaches the average density at large distances faster than $|x|^{-2}$.[158] Such an infrared regularization shall be understood even if not spelled out explicitly.

We can now discuss the delocalization induced by the dynamics. The kinetic term has a local effect, since it gives rise to an exponentially depressed delocalization (see Sect. 7.2). The crucial term is the interaction and its effect in the case of long range potential can be displayed in the limit in which the kinetic term is neglected (equivalently in the limit of large mass). In this limit,[159] the equation of motion

$$i \frac{d}{dt}\psi(x,t) = g \int dy \, \mathcal{V}(x-y)\, \rho(y,t)\, \psi(x,t) \tag{15.24}$$

is exactly solvable, since it implies

$$\frac{d}{dt}\rho(x,t) = 0$$

and it is therefore solved by

$$\psi(x,t) = \exp\left[-igt \int dy \, \mathcal{V}(x-y)\, \rho(y,0)\right]\psi(x,0) \equiv T(x,t)\,\psi(x,0). \tag{15.25}$$

For our purposes the relevant point is the delocalization property of the dynamics as displayed by the fall off of the field (anti)commutators at different times, which in the special case at hand is given by

$$[\,\psi(x,t),\,\psi^*(y,0)\,]_{\pm} = \mp(e^{-it\,g\mathcal{V}(x-y)} - 1)\,\psi^*(y,0)\,T(x,t)\,\psi(x,0)+$$

$$\delta(x-y)\,T(x,t), \tag{15.26}$$

where the $\pm$ refers to the fermion/boson case, respectively. Thus, for large space separations the r.h.s. decreases like $t\,\mathcal{V}(x-y)$, i.e. the dynamical delocalization of the (anti)commutators of the canonical variables is given by the range of the interaction potential.

On the basis of the above result, Swieca argues that such a connection between the dynamical delocalization and the range of the potential should remain valid also when one takes into account both the interaction term and the kinetic term, since the latter one by itself induces an exponentially decreasing delocalization and should therefore essentially maintain the delocalization induced by the former (*Swieca's argument*).

---

[158] This means that the class of infrared regular states ω must have the property that their correlation functions $|x|^{-1}\omega(A\,\rho(x)\,B)$, A, B any polynomials in the fields ψ, ψ^*, are absolutely integrable in x.

[159] We follow J.A. Swieca, Comm. Math. Phys. **4**, 1 (1967).

Swieca's argument can be further supported by a simple computation using Zassenhaus' formula[160]

$$e^{\lambda(A+B)} = e^{\lambda A}\, e^{\lambda B}\, e^{\lambda^2 C_2}\, e^{\lambda^3 C_3}\ldots$$

where the operators C_n are computed recursively

$$C_2 = -\tfrac{1}{2}[A,\, B],\quad C_3 = \tfrac{1}{3}[B,\, [A,\, B]] + \tfrac{1}{6}[A,\, [A,\, B]],\text{ etc.}$$

By applying the formula to the evolution operator $H = H_0 + H_1$, one gets

$$e^{-it(H_0+H_1)} = e^{-itH_1}\, e^{-itH_0}\, e^{t^2\,\frac{1}{2}\,[H_1,\,H_0]}\, e^{-it^3\,(\frac{1}{3}[H_0,\,[H_1,\,H_0]]+\frac{1}{6}[H_1,\,[H_1,\,H_0]])}\ldots$$

Now, the evolution due to the first two terms can be computed explicitly by using Swieca's results and the third term corresponds to an interaction which involves local operators and a "potential", which decreases faster than $\mathcal{V}$. In fact one has

$$i[\, H_0,\, gH_{int}\,] = \tfrac{1}{2}g \int dx\, dy\, \boldsymbol{\nabla}V(x-y)\,[\mathbf{j}(x)\rho(y) + \rho(y)\mathbf{j}(x)\,].$$

Similarly, for the fourth term one has

$$i[H_0,\, i[H_0,\, gH_{int}]] = \tfrac{1}{2}g \int dx\, dy\, \boldsymbol{\nabla}\mathcal{V}(x-y)\, \partial_t^0\,[\mathbf{j}(x)\rho(y) + \rho(y)\mathbf{j}(x)\,],$$

where ∂_t^0 denotes the derivative with respect to time of the operators with time evolution defined by the free Hamiltonian H_0 and $\mathbf{j}$ denotes (the vector part of) the current. Thus, again one has an interaction involving local operators and a "potential" $\nabla\mathcal{V}$. Moreover, for the second term in C_3, one has

$$[H_{int},\, [H_0,\, H_{int}]] = \int dx\, dy\, \nabla_k\mathcal{V}(x-y)\,\nabla_k\rho(x)\,\rho(y) \int dz\, \mathcal{V}(x-z)\rho(z)$$

$$+2\int dx\, dy\, \nabla_k\mathcal{V}(x-y)(\psi^*\psi)(x)\rho(y) \int dz\nabla_k\mathcal{V}(x-z)\rho(z).$$

Again, one has the derivative of the potential.

In any case, the effect of such terms on the evolution of the field operators gives rise to contributions to the field (anti)commutators at different times with faster decrease than that of $\mathcal{V}$ because they involve derivatives of the potential or of the fields.

The same conclusions are reached if one expands the time evolution of the fields in powers of t. To each order, the leading contribution to the large distance delocalization of the commutator is given by the potential; all other

[160] W. Magnus, Commun. Pure Appl. Math. **7**, 649 (1954); R.M. Wilcox, J. Math. Phys. **8**, 962 (1967).

terms involve derivatives of $\mathcal{V}$. For example, one has

$$\psi(x,t) = \psi(x,0) + i\,t\,(\Delta - g\,\mathcal{V} * \rho)\,\psi(x,0) +$$

$$-\tfrac{1}{2}t^2[(\Delta - g\mathcal{V} * \rho)(\Delta - g\mathcal{V} * \rho) - ig\boldsymbol{\nabla}\mathcal{V} * \mathbf{j}]\,\psi(x,0) + \dots$$

where all the functions inside the square bracket are computed at the space point x and at zero time.

Then, when one takes the (anti)commutator with $\psi^*(y,0)$, most of the terms contain derivatives of $\mathcal{V}$ and the large distance decay is governed by the fall off of the interaction potential $\mathcal{V}$.

The above arguments indicate that the large distance delocalization of the field (anti)commutators is given by the decay of the two-body potential and, therefore, Swieca's condition is satisfied if in s dimensions the potential decreases faster than $|x|^{1-s}$. In three dimensions this would imply that $\mathcal{V}(x) \sim |x|^{-2}$ is the critical decay.

Actually, since Swieca's condition is relevant for estimating the right hand side of (15.6) and typically the current, being proportional to the momentum density, involves derivatives of the fields, the critical decay may turn out to be one power slower. In fact, this mechanism is clearly displayed in the approximation in which the kinetic term is neglected (see above); the current density at time t is

$$j_k(x,t) = j_k(x,0) - 2\,t\,g\,(\psi^*\,((\nabla_k\mathcal{V}) * \rho)\,\psi)(x,0)$$

and the commutator with $\psi^*(y,0)$ is

$$[j_k(x,t),\,\psi^*(y,0)] = -2t\,g\,(\psi^*\,(\nabla_k\mathcal{V}) * \rho)(x,0)\,\delta(x-y) + \text{local terms}.$$

Thus, the delocalization is given by the derivative of the potential.

The same conclusion is reached by expanding the time evolution induced by the full Hamiltonian in powers of t. To first order in t one has

$$j_k(x,t) = j_k(x,0) + t[(\Delta\nabla_k\psi^*)\psi + \psi^*\Delta\nabla_k\psi - g\,\psi^*\,((\nabla_k\mathcal{V}) * \rho)\,\psi](x,0) -$$

$$-g\,t\,\nabla_k[\psi^*\,(\mathcal{V} * \rho)\,\psi](x,0) + O(t^2).$$

Thus, apart from local terms, the contributions to the large distance delocalization of the field (anti)commutators $[\,j_k(x,t), \psi(y,0)\,]$ either decrease as the derivative of the potential or like $\nabla_k[\rho(x,0)\mathcal{V}(x)]$, i.e. faster than the potential, since $\mathcal{V}(x)\rho(x)$ must be absolutely integrable as a regularity condition on the states for the removal of the infrared cutoff (see the above discussion).

It may be interesting to note that in the above class of models the charge integrability condition is satisfied even in the presence of long range potentials. In fact, in the approximation in which the kinetic term is neglected (see the above discussion), one has

$$\rho(x,t) = \rho(x,0)$$

and the property follows from the equal time (anti)commutators.

The same conclusion is reached by expanding the time evolution, induced by the full Hamiltonian, in powers of t. For example, to order t^2 one has

$$\rho(x,t) - e^{it\,H_0}\rho(x,0)e^{-itH_0} = \tfrac{1}{2}t^2\,g\int dy\,\nabla_k\left(\nabla_k \mathcal{V}(x-y)\rho(x,0)\,\rho(y,0)\right),$$
(15.27)

which combines the faster decrease of the derivatives of the potential with the vanishing of $\rho(x)$ at large distances faster than $|x|^{-2}$; this is the regularity condition on the states (mentioned in footnote 158), which is needed for the removal of the infrared cutoff in the equations of motion.

15.3 The Goldstone Theorem with Mathematical Flavor

After the critical discussion of the hypotheses we revisit the simple proof of Sect. 15.1 with mathematical care, also because the usual proofs for non-relativistic systems do not have the same level of rigor and sharpness as in the relativistic case.

Theorem 15.3. *(Non-relativistic Goldstone Theorem)*[161] *If*

I. $\beta^\lambda, \lambda \in \mathbf{R}$ *is a one-parameter internal symmetry group, i.e.*

$$[\beta^\lambda, \alpha_{\mathbf{x}}] = 0, \quad [\beta^\lambda, \alpha_t] = 0, \quad \forall \lambda \in \mathbf{R}, \mathbf{x} \in \mathbf{R}^s, t \in \mathbf{R},$$
(15.28)

II.   *on a subalgebra $\mathcal{A}_0$ of $\mathcal{A}$, stable under time evolution, β^λ is locally generated by a charge in the sense of (15.1,15.3-4), with Q_R defined by (15.12-13) and satisfying the charge integrability condition (15.14),*

III. β^λ *is spontaneously broken in a representation π defined by a translationally invariant ground state Ψ_0, in the sense that there exists a (self-adjoint) $A \in \mathcal{A}_0$ such that*

$$< \delta A >_0 = i \lim_{R \to \infty} < [Q_R, A] >_0 = b \neq 0,$$
(15.29)

*then, in such a representation, there exist quasi particle excitations with infinite lifetime in the limit $\mathbf{k} \to 0$ and with energy $\omega(\mathbf{k}) \to 0$ as $\mathbf{k} \to 0$ (**Goldstone quasi particles**). The corresponding states have non-trivial components in the subspaces $\{\pi(\alpha_t(A))\Psi_0\}$, $\{\pi(Q_R)\Psi_0\}$.*

Remark 1. To avoid distributional problems, it is convenient to consider a regularized charge density commutator (for simplicity the boldface notations for vectors in $\mathbf{R}^s$ is omitted and $j_0(f_x) \equiv \int dy\,f_x(y)\,j_0(y,0)$)

$$J_f(x,t) \equiv i < [\,j_0(f_x), \alpha_{-t}(A)\,] >_0\,,$$
(15.30)

with $f_x(y) = f(x+y) \in \mathcal{S}_{real}(\mathbf{R}^s)$, $\int f(y)dy = 1$. By the integrability of the

---

[161] G. Morchio and F. Strocchi, J. Math. Phys. **28**, 622 (1987).

charge density commutators, one has

$$\int dx\, J_f(x,t) = \int dx\, dy\, f(x+y) J(y,t) = \int dy\, J(y,t),$$

as a distribution in t. Moreover, as a distribution in t, J_f is absolutely integrable in x

$$\int dx |J_f(x,t)| \le \int dx\, dy |f(x+y)|\, |J(y,t)| = \int dz\, |f(z)| \int dy\, |J(y,t)|.$$

Thus

$$J(t) \equiv i \lim_{R\to\infty} < [j_0(f_R),\, \alpha_{-t}(A)] > = \int dy\, J(y,t) = \int dx\, J_f(x,t).$$

Remark 2. From a physical as well as a mathematical point of view, the issue is the relation between the limit $R \to \infty$ and the zero momentum limit of the energy spectrum. In fact, the time independence of $\lim_{R\to\infty} < [Q_R(t), A] >_0$ implies that its Fourier transform is proportional to $\delta(\omega)$; one has to prove that the point $\omega = 0$ arises from states orthogonal to the ground state and it is the limit of the energy spectrum when $k \to 0$. This is essentially guaranteed by the charge integrability condition which ensures that $\tilde{J}(k,t)$ is a continuous function of k, so that the limit $R \to \infty$, which corresponds to the limit $k \to 0$ of $\tilde{J}(k,t)$, is related to the continuous limit of the energy spectral support on real symmetric test functions $\tilde{g}(\omega) = \bar{\tilde{g}}(\omega) = \tilde{g}(-\omega)$

$$\lim_{k\to 0} \tilde{J}(k,\omega) = -2(2\pi)^2 \lim_{k\to 0} \text{Im} < j_0(f)\Psi_0,\, dE(\omega)\, dE(k)\, A\Psi_0 > .$$

The charge integrability (condition) settles the problem of the possible noncontinuity of $\tilde{J}(k,\omega)$ at $k \to 0$, raised by Klein and Lee[162] as a mechanism for evading the Goldstone theorem and accounting for an energy gap associated to symmetry breaking. The recourse to (approximate) locality to guarantee analyticity in k, as advocated by Kibble and collaborators,[163] isolates a much too strong condition, which in particular is not satisfied by systems with long range dynamics, whereas, as discussed in Sect. 15.2, the integrability of the charge density commutators seems general enough.

Proof. By the spectral theorem applied to $U(x)\, U(-t)$, one has that $\tilde{J}_f(k,\omega)$ is a measure in k and ω and by the regularity of f it is a finite measure in k. Furthermore, by the absolute integrability (in x) of $J_f(x,t)$, one has that, as

[162] A. Klein and B.W Lee, Phys. Rev. Lett. **12**, 266 (1964).
[163] T.W.B. Kibble, Broken Symmetries, in *Proc. Oxford Internat. Conf. on Elementary Particles*, Oxford 1965, p.19; G.S. Guralnik, C.R. Hagen and T.W.B. Kibble, Broken Symmetries and the Goldstone Theorem, in *Advances in Particle Physics*, Vol 2., R.L. Cool, R.E. Marshak eds., Interscience 1968.

a distribution in ω, $\tilde{J}_f(k, \omega)$ is continuous in k as $k \to 0$. Thus

$$\tilde{J}(\omega) = \lim_{k \to 0} \tilde{J}(k, \omega) = \tilde{J}(0, \omega).$$

By definition $J(t)$ is real, so that $\tilde{J}(\omega) = \overline{\tilde{J}}(-\omega)$ vanishes on test functions $\tilde{g}(\omega) = -\bar{\tilde{g}}(-\omega)$, whereas if $\tilde{g}(\omega) = \bar{\tilde{g}}(-\omega)$ one has

$$\tilde{J}(\tilde{g}) = (2\pi)^2 \lim_{k \to 0} i \int d\omega\, \tilde{g}(\omega)[< j_0(f)\Psi_0,\, dE(-\omega)\, dE(k)\, A\Psi_0 >$$

$$-\overline{< j_0(f)\Psi_0,\, dE(\omega)\, dE(-k)\, A\Psi_0 >}]$$

$$= -2(2\pi)^2 \mathrm{Im} \int d\omega\, \tilde{g}(\omega)\, < j_0(f)\Psi_0,\, dE(-\omega)\, dE(k = 0)\, A\Psi_0 > .$$

Thus, as a distribution on real symmetric test functions

$$\tilde{J}(\omega) = -2(2\pi)^2 \mathrm{Im} < j_0(f)\Psi_0,\, dE(-\omega)\, dE(k = 0)\, A\Psi_0 > . \qquad (15.31)$$

In conclusion, in the limit $k \to 0$ the imaginary part of the matrix elements of the energy spectral projection between the states $j_0(f)\Psi_0$ and $A\Psi_0$ is given by $\tilde{J}(\omega)$. As in Sect. 15.1, the stability under time evolution of the algebra $\mathcal{A}_0$ on which β^λ is generated by Q_R implies that $J(t)$ is independent of time so that $\tilde{J}(\omega) \sim \delta(\omega)$; moreover, by the above argument, $\omega = 0$ is the limit of the energy spectral support when $k \to 0$.

The ground state cannot contribute to $\tilde{J}(\omega)$ since, for real symmetric test functions $\tilde{g}(\omega)$, $\tilde{h}(k)$, $< j_0(f)dE(\tilde{g})\, dE(\tilde{h}) >_0 < A >_0$ is real. The infinite lifetime is implied by the continuity in k of the energy spectrum, which shrinks to zero when $k \to 0$.

The above version of the Goldstone theorem improves the standard treatment (for non-relativistic systems) in i) identifying the relevant hypotheses, in a way which looks applicable to the physically interesting cases, ii) emphasizing the role of the localization properties of the dynamics, and iii) predicting the existence of quasi particle Goldstone bosons.

As we shall see, the existence of stable Goldstone particles is related to relativistic locality and spectrum.

16 * The Goldstone Theorem at Non-zero Temperature

The proof of the Goldstone theorem can be easily extended to the case of non-zero temperature $T = 1/\beta$, i.e. to representations defined by KMS states. In this case, the interest of the theorem is in the prediction of the Goldstone quasi particles, and the derivation of such a prediction crucially depends on the integrability of the charge density commutators. The absence of an energy gap (as in Theorem 15.1) is not very significant, since it is already implied by general properties (like timelike clustering) of the KMS states.[164]

As we shall see, in the non-zero temperature case the fine mathematical points discussed in Sect. 15.3, e.g. the distributional properties of $J(x,t)$, become more relevant also in view of some puzzling statements that have appeared in the literature.

Theorem 16.1. *(Non-relativistic Goldstone Theorem for $T \neq 0$)*
Under the assumptions I, II, III of Theorem 15.3, with π a representation defined by a translationallly invariant KMS state Ω, the same conclusions hold (existence **Goldstone quasi particles***).*

Moreover, if $< >$ denotes the expectation on Ω, the Fourier transform $\tilde{\mathcal{W}}(k,\omega)$ of the two point function

$$\mathcal{W}(x,t) \equiv < j_0(f_x)\,\alpha_t(A) >,$$

satisfies (with b defined in (15.29))

$$i \lim_{k \to 0} [\,\tilde{\mathcal{W}}(k,\omega) - \overline{\tilde{\mathcal{W}}(-k,-\omega)}\,] = b\,\delta(\omega), \tag{16.1}$$

---

[164] R. Haag, D. Kastler and E.B. Trych-Pohlmeyer, Comm. Math. Phys. **38**,137 (1974), Prop. 3. As a consequence of this result, several papers have been devoted to a version of the Goldstone theorem which relates symmetry breaking to poor clustering, rather than to the absence of energy gap: L. Landau, J. Fernando Perez and W.F. Wreszinski, J. Stat. Phys. **26**, 755 (1981); Ph. Martin, Nuovo Cim. **68**, 302 (1982); M. Fannes, J.V. Pulé and A. Verbeure, Lett. Math. Phys. **6**, 385 (1982) and the reviews Ch. Gruber and P.A. Martin, Goldstone theorem in Statistical mechanics, in *Mathematical Problems in Theoretical Physics*, (Berlin Conference 1981), R. Schrader et al. eds., Springer 1982, p. 25; W.F. Wreszinski, Fortschr. Phys. **35**, 379 (1987).

F. Strocchi: * The Goldstone Theorem at Non-zero Temperature, Lect. Notes Phys. **732**, 177–179 (2008)
DOI 10.1007/978-3-540-73593-9_16   © Springer-Verlag Berlin Heidelberg 2008

as a distribution in ω, and

$$\lim_{k \to 0} [\tilde{\mathcal{W}}(k, \omega) + \overline{\tilde{\mathcal{W}}(-k, -\omega)}] = 2i\,(b/\beta)\,\delta'(\omega) \qquad (16.2)$$

as a distribution in ω on test functions $g(\omega) \in (1 - e^{-\beta\omega})\,\mathcal{S}(\mathbf{R})$, in particular on antisymmetric test functions of compact support.

Proof. The first part follows as in Theorem 15.3; by the same arguments one has

$$\tilde{J}(\omega) \equiv \lim_{k \to 0} \tilde{J}(k, \omega) = i \lim_{k \to 0} [\tilde{\mathcal{W}}(k, \omega) - \overline{\tilde{\mathcal{W}}(-k, -\omega)}] =$$

$$\lim_{k \to 0} -2(2\pi)^2 \mathrm{Im} < j_0(f)\,dE(\omega)\,dE(k)\,A >= b\,\delta(\omega), \qquad (16.3)$$

as a distribution in ω.

Now, the KMS condition gives

$$\tilde{J}(k, \omega) = i\,(1 - e^{-\beta\omega})\tilde{\mathcal{W}}(k, \omega) \qquad (16.4)$$

and, by the reality of $J(x, t)$,

$$\tilde{J}(k, \omega) = \overline{\tilde{J}(-k, -\omega)} = -i\,(1 - e^{\beta\omega})\overline{\tilde{\mathcal{W}}(-k, -\omega)}. \qquad (16.5)$$

By adding (16.4) to (16.5) times $e^{-\beta\omega}$, one gets

$$(1 + e^{-\beta\omega})\,\tilde{J}(k, \omega) = i\,(1 - e^{-\beta\omega})\,[\tilde{\mathcal{W}}(k, \omega) + \overline{\tilde{\mathcal{W}}(-k, -\omega)}]. \qquad (16.6)$$

The charge (commutator) integrability condition implies that the right hand side is a continuous function of k, as a distribution in ω, so that, on test functions $g(\omega) \in (1 - e^{-\beta\omega})\,\mathcal{S}(\mathbf{R})$, also the term in square brackets on the r.h.s. of (16.6) has a limit for $k \to 0$.

Then (16.3), (16.6) imply (16.2). Clearly $(1 - e^{-\beta\omega})\,\mathcal{S}(\mathbf{R})$ contains all the antisymmetric test functions of compact support.

The occurrence of the δ' should not appear strange; by the unitarity of the space and time translations $\tilde{\mathcal{W}}(k, \omega)$ is a measure in (k, ω), but in general it is not a measure in ω for k fixed; in particular, it need not to be a measure in the limit $k \to 0$. By the charge integrability condition $\tilde{J}(k, \omega)$ and therefore $(1 - e^{-\beta\omega})\,\tilde{\mathcal{W}}(k, \omega)$ is a continuous function of k as a distribution in ω, but this does not imply that $\tilde{\mathcal{W}}(k, \omega)$ is a continuous function of k as a measure in ω and in particular that it is a measure in ω in the limit $k \to 0$.[165]

---

[165] Such an incorrect implication is at the basis of no-go theorems about spontaneous symmetry breaking at non-zero temperature as in R. Requardt, J. Phys. A: Math. Gen. **13**, 1769 (1980), Theorem 1. For a discussion of these problems and its relevance for the derivation of the f-sum rule and of the long-wavelength "perfect screening" sum rule, see G. Morchio and F. Strocchi, Ann. Phys. **185**, 241 (1988); Errata **191**, 400 (1989) ; there one can also find a detailed discussion of the case $j_0(x) = \rho(x)$, $A = \dot{\rho}(x)$.

Such a continuity in k as a measure in ω would hold if the charge density commutator is absolutely integrable, uniformly in time.

These mathematical delicate points are clearly displayed by the free Bose gas or by the massless scalar field and it is instructive to work out these applications of the general statements.

For example, for a massless scalar field $\varphi(x,t)$ at non-zero temperature $T = 1/\beta$, the charge density $\partial_0\varphi$, associated with the conserved current $\partial_\mu\varphi$, generates the spontaneously broken symmetry: $\varphi \to \varphi + \lambda$. According to the discussion of Chap. 14, one can compute the two point function $< \dot{\varphi}(x,t)$ $\varphi(y,t') >$ and its Fourier transform $-i\omega\tilde{W}(k,\omega)$, getting

$$-i\omega\tilde{W}(k,\omega) = \tfrac{1}{2}i\,(1 - e^{-\beta|\omega|})^{-1}\,[\,\delta(\omega - |k|) - e^{-\beta|k|}\,\delta(\omega + |k|)\,].$$

It is continuous in k as a distribution in ω, but it is easy to see that it is not a measure in ω in the limit $k \to 0$.[166] It is also clear that the charge (commutator) integrability condition holds as a distribution in the time variable.

Similar features are shared by the gauge symmetry breaking in the free Bose gas for $T < T_c$. The expectations $\omega_\theta([j_0(x), \alpha_t(a(h))])$, $h \in \mathcal{S}$, with $j_0(x) = \psi^*(x)\psi(x)$ are absolutely integrable in x.[167] However, even if one uses a subtracted density $j_0^s(x) \equiv j_0(x) - \omega_\theta(j_0(x))$, the two point function $W(x,t) \equiv \omega_\theta(j_0^s(x)\,\alpha_t(a(h)))$ is not integrable in x.[168]

---

[166] In fact, after smearing with a test function $g(\omega)$, one has

$$\tilde{W}(k, -i\omega g) \equiv \int d\omega\,\tilde{W}(k,\omega)(-i\omega)\,g(\omega) =$$

$$= \tfrac{1}{2}i(1 - e^{-\beta\,|k|})^{-1}[g(|k|) - e^{-\beta\,|k|}\,g(-|k|)\,]$$

$$\sim_{k\to 0} \tfrac{1}{2}i[\,g(0) + |k|\,g'(0)\,(1 + e^{-\beta\,|k|})/(1 - e^{-\beta\,|k|})\,],$$

so that , $\tilde{W}(0, -i\omega g) = (i/2)\,[\,g(0) + 2g'(0)/\beta\,]$. Thus, $-i\omega\tilde{W}(k,\omega)$ is not a measure in ω in the limit $k \to 0$.

[167] By the CCR and (13.20), such expectations are proportional to the Fourier transform of $h(k)\exp ik^2 t \in \mathcal{S}$.

[168] By using (13.14), (13.15) and the quasi free property of ω_θ, one has

$$\omega_\theta((a^*(q)\,a(q') - <a^*(q)\,a(q') >)\,a(p)) = \rho_0^{1/2}\,e^{i\theta}\,\delta(q - p)\,\delta(q')\,(e^{\beta q^2} - 1)^{-1}.$$

Therefore,

$$\tilde{W}(k,\omega) = \rho_0^{1/2}e^{i\theta}\,h(k)\,\delta(\omega - k^2)\,(e^{\beta\,\omega} - 1)^{-1},$$

which is not continuous in k (not even as a distribution in ω).

17 The Goldstone Theorem
for Relativistic Local Fields

Relativistic systems, like elementary particles, are described by an algebra of observables $\mathcal{A}_{obs}$ which satisfies the causality condition, (4.2), and is stable under the automorphisms $\alpha(a, \Lambda)$ which describe space time translations and Lorentz transformations (with parameters a, Λ, respectively).

The physically relevant representations of $\mathcal{A}_{obs}$ have to satisfy the relativistic version of conditions I-III (Chap. 5):

I. (**Poincaré Covariance**) The automorphisms $\alpha(a, \Lambda)$ are implemented by a strongly continuous group of unitary operators $U(a, \Lambda)$.

II. (**Relativistic spectral condition**)

$$H \geq 0, \quad H^2 - \mathbf{P}^2 \geq 0,$$

equivalently, the Fourier transform of the matrix elements of $U(a)$ have support in the *closed forward cone* $\overline{V}_+ = \{p^2 \geq 0,\, p_0 \geq 0\}$.

III. (**Vacuum state**) There is a unique space-time translationally invariant state Ψ_0 (vacuum state) cyclic for the algebra $\mathcal{A}_{obs}$.

As we have also seen in the case of non-relativistic systems (e.g. the free Bose gas), it is convenient (if not necessary) for the formulation and solution of the dynamical problem to work with an extension of the algebra of observables. This amounts to introducing a field algebra $\mathcal{F}$, which plays the role of the algebra of canonical variables of the non-relativistic systems.

In quantum field theory, the algebra $\mathcal{F}$ is generated by a set of *fields* $\{\varphi_j(x),\, x = (\mathbf{x}, x_0),\, j \in I = \text{finite index set}\}$, which are operator valued (tempered) distributions and in general transform covariantly under the Poincaré group

$$U(a, \Lambda(A))\, \varphi_j(x)\, U(a, \Lambda(A))^{-1} = S_{jk}(A^{-1})\, \varphi_k(\Lambda(A)x + a), \quad A \in SL(2, \mathbf{C}),$$

$$(17.1)$$

where S_{jk} is a finite dimensional representation of $SL(2, \mathbf{C})$ (the universal covering of the restricted Lorentz group $L_+^\uparrow$).

F. Strocchi: *The Goldstone Theorem for Relativistic Local Fields*, Lect. Notes Phys. **732**, 181–188 (2008)
DOI 10.1007/978-3-540-73593-9_17

For example, for a scalar field φ, $S_{jk} = 1$ and for a vector field j_μ, $S_{\mu\nu}(\Lambda^{-1}) = (\Lambda^{-1})^\nu_\mu$, etc.

The construction of a C^*-algebra $\mathcal{A}_\mathcal{F} \supset \mathcal{A}_{obs}$, in terms of the polynomial algebra $\mathcal{F}$ generated by the smeared fields $\{\varphi_j(f), f \in \mathcal{S}(\mathbf{R}^4)\}$, requires self-adjoint conditions on the smeared fields, which are not easy to control (in contrast with the finite dimensional case). Therefore, following Wightman[169] and also in analogy with the perturbative approach to quantum field theory, one usually works directly with the (polynomial) field algebra $\mathcal{F}$.

In general, it is not automatic that the field algebra $\mathcal{F}$, needed for the formulation and solution of the dynamical problem, is a local algebra, i.e. it satisfies (4.2) or its extension for anticommuting fields.

For example, this is not the case of the Coulomb gauge field algebra of quantum electrodynamics (QED), where the electron field ψ and the electromagnetic field $F_{\mu\nu}$ do not commute at spacelike separations; moreover, ψ and the vector potential do not transform as in (17.1).

On the other hand, a local covariant field algebra $\mathcal{F}$ is at the basis of the so-called renormalizable gauges of gauge quantum field theory (e.g. the Feynman gauge in QED), at the price that the vacuum state is not a positive functional on $\mathcal{F}$.[170]

Even in this more general case without positivity[171], one can introduce the concept of symmetry breaking of a one-parameter group of automorphisms β^λ of $\mathcal{F}$, when the vacuum expectation values (v.e.v.) of the fields, briefly denoted by $< A >_0$, $\forall A \in \mathcal{F}$, are not invariant under β^λ, i.e. $< \delta A >_0 \neq 0$, for some $A \in \mathcal{F}$. One may then investigate the consequences of such a breaking for the spectral support of the Fourier transforms of the v.e.v.

We shall now discuss a version of the Goldstone theorem, which applies to local field algebras with v.e.v. which satisfy space-time translation invariance, relativistic spectral support, but not necessarily positivity.[172]

---

[169] R.F. Streater and A.S. Wightman, *PCT, Spin and Statistics and All That*, Benjamin-Cummings 1980.

[170] For a general discussion of the interplay between locality and positivity in gauge quantum field theory, see F. Strocchi, *Selected Topics on the General Properties of Quantum Field Theory*, World Scientific 1993 and references therein.

[171] For the discussion of this more general formulation of quantum field theories, which is particularly relevant for two-dimensional models involving a massless scalar field and for gauge quantum field theories, see F. Strocchi and A.S. Wightman, J. Math. Phys. **15**, 2198 (1974); G. Morchio and F. Strocchi, Ann. Inst. H. Poincaré, **A 33**, 251 (1980) and for a general review F. Strocchi, *Selected Topics on the General Properties of Quantum Field Theory*, World Scientific 1993, Chap. VI.

[172] F. Strocchi, Comm. Math. Phys. **56**, 57 (1977).

Theorem 17.1. *(Goldstone Theorem for relativistic local fields) Let β^λ be a one-parameter group of *-automorphisms of the field algebra $\mathcal{F}$, which*

I. *commutes with space-time translations,*
II. *is locally generated by a charge, in the sense that there is a local covariant conserved current j_μ such that $\forall A \in \mathcal{F}$*

$$\delta A = i \lim_{R \to \infty} [Q_R, A],$$

$$Q_R \equiv j_0(f_R, \alpha) \equiv \int d^4x \, f_R(\mathbf{x}) \, \alpha(x_0) \, j_0(\mathbf{x}, x_0), \qquad (17.2)$$

with f_R as in (15.12), and

$$\alpha \in \mathcal{D}(\mathbf{R}), \; supp \, \alpha \subseteq [-\delta, \delta], \; \tilde{\alpha}(0) = \int dx_0 \, \alpha(x_0) = 1, \qquad (17.3)$$

III. *is spontaneously broken in the sense that there exists at least one $A \in \mathcal{F}$ with $< \delta A >_0 \neq 0$.*

*Then, the Fourier transform of the two point function $< j_0(x) A >$ contains a $\delta(p^2)$ singularity (**Goldstone massless modes**).*

Remark 1. Relativistic local fields are more singular than non-relativistic fields and therefore a smearing in time is necessary to get mathematically well defined objects;[173] this is the reason for the introduction of the test function $\alpha(x_0)$ and $\tilde{\alpha}(0) = 1$ is merely a normalization condition. Indeed, even for a free Dirac field, $j_\mu(\mathbf{x}, x_0)$ is a distribution in the four variables, which does not admit a restriction at fixed time; in fact the commutator $[j_0(\mathbf{x}, x_0), j_i(\mathbf{y}, x_0)]$ is a divergent Schwinger term.[174] However, the introduction of the smearing with α does not spoil the simple meaning of condition II and its possible control, thanks to the following Lemma.

Remark 2. For the symmetry breaking condition it is enough to consider the case in which A is localized in a bounded space time region, briefly $A \in \mathcal{F}_{loc}$, since $< \delta A >_0 = 0$ for all such A implies the invariance for all $A \in \mathcal{F}$ by a density argument.

Lemma 17.2. *As a consequence of locality, for any $A \in \mathcal{F}$ the limit $R \to \infty$ in (17.2) exists and it is independent of α (with $\tilde{\alpha}(0) = 1$).*

Proof. In fact, if A is a local field with compact support K, the commutator $[j_\mu(\mathbf{x}, \alpha), A]$ vanishes by locality for $|\mathbf{x}|$ sufficiently large and for a general

---

[173] A.S. Wightman, Ann. Inst. H. Poincaré, I, 403 (1964).
[174] For a simple discussion see e.g. F. Strocchi, *Selected Topics on the General Properties of Quantum Field Theory*, World Scientific 1993, Sect. 4.5.

$A \in \mathcal{F}$ the commutator decreases faster than any inverse power of $|\mathbf{x}|$. Therefore, the integrability of the charge commutators is automatically satisfied (the local integrability is not a problem as discussed in Sect. 15.2).

Moreover, if α_1, $\alpha_2 \in \mathcal{D}(\mathbf{R})$ are two normalized test functions, then

$$\alpha_1 - \alpha_2 = d\beta/dx_0, \quad \beta(x_0) \equiv \int_{-\infty}^{x_0} dx_0' \left(\alpha_1(x_0') - \alpha_2(x_0')\right) \in \mathcal{D}(\mathbf{R}),$$

and by current conservation, $\partial^\mu j_\mu = 0$,

$$[j_0(f_R, \alpha_1) - j_0(f_R, \alpha_2), A] = -[\partial_0 j_0(f_R, \beta), A] = [\mathbf{j}(\boldsymbol{\nabla} f_R, \beta), A].$$

Since supp $\boldsymbol{\nabla} f_R \subseteq \{R \le |\mathbf{x}| \le R(1+\varepsilon)\}$, for R large enough, the localization region of $\mathbf{j}(\boldsymbol{\nabla} f_R, \beta)$ becomes spacelike with respect to any (bounded) compact set K and the commutator vanishes by locality.

Remark 3. The argument of the Lemma can be adapted to the case in which A is replaced by a local field variable, say $\varphi(\mathbf{y}, y_0)$, since $[j_0(f_R, \alpha), \varphi(\mathbf{y}, y_0)]$ is a well defined operator valued distribution in $\mathbf{y}, y_0$ and by locality the limit $R \to \infty$ exists and it is actually reached for R large enough. By the same argument as above, such a limit is independent of α and therefore taking $\alpha_1(x_0) = \alpha(x_0 - y_0)$, with α as in (17.3), the limit of shrinking support $\delta \to 0$ exists and defines a regularized version of the equal time commutator between $j_0(f_R, x_0)$ and $\varphi(\mathbf{y}, x_0)$, for R large enough.[175]

Remark 4. As a consequence of the above Lemma, the delicate problems of the non-relativistic case (discussed in Sect. 15.2) do not arise for local field algebras. By Remark 3, the existence and identification of a (conserved) current which generates a given (algebraic) symmetry β^λ can be inferred by using the (equal time) CCR (or ACR) and the stability under time evolution is guaranteed by the independence of α.

The proof of the theorem is particularly simple if the order parameter is given by a local field, say $\varphi(\mathbf{y}, y_0)$, which transforms as in (17.1), briefly called an *elementary field*. For a generic element $A \in \mathcal{F}$, one can easily obtain covariance under space time translations by putting $A_x = \alpha_x(A)$, but then the transformation under the Poincaré group is not given by (17.1). The Lorentz invariance of the v.e.v. requires that the order parameter is a scalar and thus one may take φ a scalar. This is the case considered in the classic work of Goldstone, Salam and Weinberg,[176] which we reproduce below in a somewhat simplified version.

---

[175] The effectiveness of such a regularization in giving finite results is clearly displayed by the equal time commutator $[j_0(f_R, x_0), j_i(\mathbf{y}, x_0)]$, for a free Dirac current.

[176] J. Goldstone, A. Salam and S. Weinberg, Phys. Rev. **127**, 965 (1962).

Proof for elementary fields. The Poincaré covariance implies that

$$J_\mu(x-y) \equiv\; < j_\mu(x)\, \varphi(y) >_0 = (\Lambda^{-1})^\nu_\mu J_\nu(\Lambda(x-y))$$

and therefore by a general result[177]

$$J_\mu(x) = \partial_\mu F(x), \quad F(x) = F(\Lambda x). \qquad (17.4)$$

Now, current conservation implies

$$\Box F(x) = 0,$$

so that the Fourier transform is of the form $\tilde{F}(p) = f(p)\delta(p^2)$. Finally, the symmetry breaking condition III excludes $f(p) = 0$.

The Goldstone-Salam-Weinberg (GSW) version of the Goldstone theorem does not cover the case in which the symmetry breaking involves a polynomial of the fields (or a composite field). For these reasons a more general version is important.[178]

---

[177] K. Hepp, Helv. Phys. Acta **36**, 355 (1963). The proof of (17.4) can be reduced to an exercise in relativistic kinematics. By Poincaré covariance the Fourier transform $J_\mu(p)$ satisfies $J_\mu(p) = (\Lambda^{-1})^\nu_\mu J_\nu(\Lambda p)$ and therefore if $\mathbf{q} = R\mathbf{p}$, R a rotation,

$$|\mathbf{p}|^2 J_i(p_0, \mathbf{p}) - p_i p_k J_k(p_0, \mathbf{p}) = |\mathbf{q}|^2 (R^{-1} J)_i(p_0, \mathbf{q}) - (R^{-1} q)_i q_k J_k(p_0, \mathbf{q}).$$

The l.h.s. vanishes for $\mathbf{p}$ pointing in the i-direction and therefore, multiplying the r.h.s. by R, for any $\mathbf{q} \neq 0$, $J_i(p_0, \mathbf{q}) = q_i\, \mathbf{q} \cdot \mathbf{J}(p_0, \mathbf{q})/|\mathbf{q}|^2$. Again, by using rotation covariance, (omitting the variable p_0),

$$G(\mathbf{p}) \equiv \mathbf{p} \cdot \mathbf{J}(\mathbf{p}) = \mathbf{p} \cdot R^{-1}\mathbf{J}(R\mathbf{p}) = R\mathbf{p} \cdot \mathbf{J}(R\mathbf{p}) = G(R\mathbf{p}),$$

i.e. $G = G(|\mathbf{p}|)$. Similarly $J_0(p) = J_0(p_0, |\mathbf{p}|)$. Moreover, by using covariance under Lorentz boosts, e.g. boosts in the 3-direction, one has $J_i(p_0, p_3, p_1, p_2) = J_i(\Lambda(p_0, p_3), p_1, p_2)$, $i = 1, 2$, i.e. they are functions of the boost invariant combination $p_0^2 - p_3^2$. Then, by rotation invariance, $G(p_0, |\mathbf{p}|)/|\mathbf{p}|^2 = F(p^2)$. Finally

$$J_3(p) = p_3 F(p^2) = ((\Lambda)^{-1})^0_3\, J_0(\Lambda p) + ((\Lambda)^{-1})^3_3\, (\Lambda p)_3\, F(p^2)$$

$$= ((\Lambda)^{-1})^0_3\, J_0(\Lambda p) + p_3 F(p^2) - ((\Lambda)^{-1})^0_3 (\Lambda p)_0 F(p^2),$$

i.e. $J_0(p) = p_0\, F(p^2)$.

[178] D. Kastler, D.W. Robinson and A. Swieca, Comm. Math. Phys. **2**, 108 (1966); H. Ezawa and J.A. Swieca, Comm. Math. Phys. **5**, 330 (1967). See also the beautiful reviews: D. Kastler, Broken Symmetries and the Goldstone Theorem, in *Proc. 1967 Int. Conf. on Particles and Fields (Rochester)*, C.R. Hagen et al. eds., Wiley 1967; J.A. Swieca, Goldstone Theorem and Related Topics, in *Chargèse Lectures* **4**, D. Kastler ed., Gordon and Breach 1969.

Such a general proof also makes clear that locality and not Lorentz covariance, as one may be led to believe on the basis of the GSW version, is the crucial ingredient. Actually, the non-covariance of the fields of the Coulomb gauge, rather than their non-locality, has been taken as explanation of the evasion of the Goldstone theorem by Higgs[179] in his proposal of the so-called Higgs mechanism. As a matter of fact, for the two point function of elementary fields, Lorentz covariance and locality are deeply related[180] and therefore it is not strange that the GSW proof, which exploits Lorentz covariance, may hide the role of locality. On the other hand, the recognition of the role of locality and its failure in positive gauges establishes a strong bridge between gauge quantum field theories and many body theories like superconductivity and Coulomb systems (see the discussion in Sect. 15.2 and in Chap. 19).

Proof of Theorem 17.1. The proof exploits the general representation of the v.e.v. of the commutator of two local fields, known as the *Jost-Lehmann-Dyson (JLD) representation*,[181] which reads

$$-i\,J(x) \equiv < [\,j_0(x),\, A]\, >_0 = i \int dm^2 \int d^3y \{\rho_1(m^2, \mathbf{y})\, \Delta(\mathbf{x} - \mathbf{y}, x_0; m^2) +$$

$$\rho_2(m^2, \mathbf{y})\, \dot\Delta(\mathbf{x} - \mathbf{y}, x_0; m^2)\}, \quad A \in \mathcal{F}_{loc}, \tag{17.5}$$

where $i\,\Delta(\mathbf{x}, x_0; m^2)$ is the commutator function $< [\varphi(x), \varphi(0)] >_0$ of a free scalar field φ of mass m. The spectral functions $\rho_i(m^2, \mathbf{y})$, $i = 1, 2$, are tempered distributions in m^2 (actually measures if positivity holds), with compact support in $\mathbf{y}$ as a consequence of locality, since the l.h.s. vanishes for $\mathbf{x}$ sufficiently large. The convolution in $\mathbf{y}$ and the integration in m^2 have to be understood as performed after smearing in x, with a test function of compact support.

The crucial ingredients for the derivation of the JLD formula are the localization properties of the commutator and the support in the forward cone of the Fourier transform of $< j_0(x)\, A >_0$, as a consequence of the spectral

---

[179] P.W. Higgs, Phys. Lett. **12**, 133 (1964).

[180] J. Bros, H. Epstein and V. Glaser, Comm. Math. Phys. **6**, 77 (1967).

[181] R. Jost and H. Lehmann, Nuovo Cim. **5**, 1598 (1957); F. Dyson, Phys. Rev. **110**, 1460 (1958); H. Araki, K. Hepp and D. Ruelle, Helv. Acta Phys. **35**, 164 (1962); A.S. Wightman, Analytic functions of several complex variables, in *Dispersion Relations and Elementary Particles*,(Les Houches Lectures), C. de Witt and R. Omnes eds., Wiley 1961; H. Araki, *Mathematical Theory of Quantum Fields*, Oxford Univ. Press 1999, Sect. 4.5.

condition. For a rigorous proof of the JLD representation we refer to the references given in the previous footnote.[182]

Now, following Ezawa and Swieca, by locality $\rho_i(m^2, \mathbf{y})$ can be written as

$$\rho_i(m^2, \mathbf{y}) = \bar{\rho}_i(m^2)\,\delta(\mathbf{y}) + \boldsymbol{\nabla} \cdot \boldsymbol{\sigma}_i(m^2, \mathbf{y}), \tag{17.6}$$

$$\bar{\rho}_i(m^2) = \int d^3y\,\rho_i(m^2, \mathbf{y}),$$

with $\boldsymbol{\sigma}_i$ of compact support in $\mathbf{y}$.[183]

By locality, the second term in (17.6) does not contribute to the charge commutator for R sufficiently large; in fact, the operator $\boldsymbol{\nabla}$ can be shifted to $\Delta(\mathbf{x} - \mathbf{y}, x_0; m^2)$ and then to $f_R(\mathbf{x})$, by partial integrations, so that the integration involves only points $\{\mathbf{x} - \mathbf{y}, x_0; |\mathbf{x}| \geq R,\ \mathbf{y} \in \operatorname{supp}\boldsymbol{\sigma}_i\}$, which are spacelike for R sufficiently large and Δ vanishes there by locality.

Thus, for R large enough,

$$< [j_0(f_R, \alpha), A] >_0 = i \int dm^2 \{\bar{\rho}_1(m^2)\,\Delta(f_R, \alpha; m^2) + \bar{\rho}_2(m^2)\dot{\Delta}(f_R, \alpha; m^2)\}$$

and

$$\Delta(f_R, \alpha; m^2) \equiv \int d^4x\,\Delta(\mathbf{x}, x_0; m^2)\,f_R(\mathbf{x})\,\alpha(x_0) =$$

$$(-i/2\pi) \int d^3p\,\tilde{f}_R(\mathbf{p})(2p_0)^{-1}[\tilde{\alpha}(p_0) - \tilde{\alpha}(-p_0)],$$

---

[182] The following heuristic argument (which does not consider the technical distributional problems) may illustrate the origin and the physical meaning of the JLD formula, (in the positive case). By inserting a complete set of improper eigenstates of the momentum $|\mathbf{p}, m^2 >$, $m^2 \equiv p^2$, $p_0 \equiv (\mathbf{p}^2 + m^2)^{1/2}$, one has (taking $A = A^*$, $j_0 = j_0^*$)

$$-i\,J(x) = \int dm^2\,d^3p/(2p_0)\,e^{i\mathbf{p}\cdot\mathbf{x}}\,[J_-(\mathbf{p}, m^2)\,\cos(p_0x_0) - iJ_+(\mathbf{p}, m^2)\,\sin(p_0x_0)],$$

$$J_\pm(\mathbf{p}, m^2) \equiv\, < j_0|\mathbf{p}, m^2 ><\mathbf{p}, m^2|A > \pm\, < A|\mathbf{p}, m^2 ><\mathbf{p}, m^2|j_0 > .$$

Since

$$\Delta(\mathbf{x}, x_0; m^2) = -(2\pi)^{-3} \int \sin(p_0x_0)e^{i\mathbf{p}\cdot\mathbf{x}}\,d^3p/p_0$$

and $\cos(p_0x_0) = p_0^{-1}\,d\,\sin(p_0x_0)/dx_0$, the integrations in d^3p give rise to convolutions, leading to (17.5), with $\rho_i(m^2, \mathbf{y})$, $i = 1, 2$, the Fourier transforms of $iJ_+(\mathbf{p}, m^2)/2$ and of $-J_-(\mathbf{p}, m^2)/2p_0$, respectively.

[183] In fact, a distribution $\rho(x) \in \mathcal{S}'(\mathbf{R})$ of compact support can be written in the form

$$\rho(x) = \delta(x) \int dy\,\rho(y) + \partial_x\sigma(x),\ \ \sigma(x) \equiv \int_{-\infty}^{x} dx'\,[\rho(x') - \delta(x') \int dy\,\rho(y)],$$

with σ of compact support. The extension to $\mathcal{S}'(\mathbf{R}^n)$ is obtained by iteratively applying the above decomposition to each variable.

$$\dot{\Delta}(f_R, \alpha; m^2) = -1/4\pi \int d^3p \, f_R(\mathbf{p}) \left[\tilde{\alpha}(p_0) + \tilde{\alpha}(-p_0) \right], \quad p_0 \equiv (\mathbf{p}^2 + m^2)^{1/2}.$$

For $\alpha(x_0)$ real and symmetric one has $\tilde{\alpha}(p_0) = \tilde{\alpha}(-p_0)$ and only the second term contributes, so that (since $f_R(\mathbf{p}) \to (2\pi)^{3/2} \, \delta(\mathbf{p})$) one has

$$\lim_{R \to \infty} < [j_0(f_R, \alpha), A] >_0 = -i\sqrt{2\pi} \int_0^\infty dm^2 \, \bar{\rho}_2(m^2) \, \tilde{\alpha}(\sqrt{m^2}). \qquad (17.7)$$

By Lemma 17.2, the r.h.s. is a functional of $\tilde{\alpha}$, which depends only on the value that $\tilde{\alpha}$ takes at the origin and therefore

$$\bar{\rho}_2(m^2) = \lambda \delta(m^2), \quad \lambda \in \mathbf{C}. \qquad (17.8)$$

The symmetry breaking condition implies $\lambda \neq 0$ and therefore the Fourier transform of two point function $< j_0(x) \, A >_0$ contains a $\delta(p^2)$.

18 An Extension of Goldstone Theorem to Non-symmetric Hamiltonians

The Goldstone theorem and its rigorous predictions on the energy spectrum at zero momentum can be extended[184] to the case in which the Hamiltonian H is not symmetric, but it has simple transformation properties, in the sense that the multiple commutators of H and the charge Q generate a finite dimensional Lie algebra, briefly

$$[Q^i, H] = c^i_k Q^k.$$

The invariance of the dynamics is then replaced by

I. (*Covariance group of the dynamics*)
There exists a Lie group G of *-automorphisms α^g, $g \in G$, of a subalgebra $\mathcal{A}_0 \subseteq \mathcal{A}$, which contains the dynamics α_t as a one-parameter subgroup; for simplicity, in the following, α^g is assumed to commute with the space translations $\alpha_{\mathbf{x}}$.

II. (*Local generation of the covariance group*)
The covariance group α^g, $g \in G$ is locally generated by charge densities

$$\delta^i A \equiv \partial \alpha^g(A)/\partial g_i|_{g=0} = i \lim_{R \to \infty} [Q^i_R, A], \quad A \in \mathcal{A}_0, \qquad (18.1)$$

$$Q^i_R(t) = \alpha^t(Q^i_R) = \int dx\, f_R(x)\, j^i_0(\mathbf{x}, t)),$$

and the charge density commutators are absolutely integrable (for large $|\mathbf{x}|$) as tempered distributions in t (the local charge generating α^t is the infrared regularized Hamiltonian H_L).[185]

Furthermore, the local charges satisfy the Lie algebra relations (as commutators on $\mathcal{A}_0$)

$$\lim_{R \to \infty} \lim_{S \to \infty} [[Q^i_S, Q^j_R], A] = \lim_{S \to \infty} \lim_{R \to \infty} [[Q^i_S, Q^j_R], A]$$

---

[184] G. Morchio and F. Strocchi, Ann. Phys. **185**, 241 (1988).

[185] As remarked in the standard case, the above commutators as well as the following ones are understood as bilinear forms on a dense set of states in each relevant representation; actually, all what is needed is their expectations on the ground state.

F. Strocchi: *An Extension of Goldstone Theorem to Non-symmetric Hamiltonians*, Lect. Notes Phys. **732**, 189–192 (2008)
DOI 10.1007/978-3-540-73593-9_18   © Springer-Verlag Berlin Heidelberg 2008

$$= \lim_{R \to \infty} c_k^{ij} [Q_R^k, A], \quad \forall A \in \mathcal{A}_0, \tag{18.2}$$

where c_k^{ij} are the structure constants of the group G.

The interchange of the order of the limits in the above equation qualifies the local generation of the group G; in particular choosing $g_1 = t$, $c_k^i \equiv c_k^{1i}$, one obtains the local covariance properties of the Hamiltonian (as commutators on $\mathcal{A}_0$)

$$\lim_{L \to \infty} \lim_{R \to \infty} [[Q_R^i, H_L], A] = \lim_{R \to \infty} \lim_{L \to \infty} [[Q_R^i, H_L], A] = \lim_{R \to \infty} c_k^i [Q_R^k, A]. \tag{18.3}$$

The following notion is relevant for the extended version of the Goldstone theorem.

Definition 18.1. *Given an $n \times n$ matrix $C = \{C_{ij}\}$, a vector J is said to have spectral support $\{\omega_1, ... \omega_k\}$, relative to C, if it is the linear combination of generalized eigenvectors of C, i.e. if one has*

$$J = \sum_{\alpha=1}^{k} a_\alpha w^\alpha, \quad a_\alpha \neq 0, \quad (C - \omega_\alpha)^{n_\alpha} w^\alpha = 0, \quad n_\alpha \in \mathbf{N}. \tag{18.4}$$

Theorem 18.2. [186] *Let G be the covariance group of the dynamics satisfying the above conditions I, II and*
III.(Symmetry breaking condition) G is spontaneously broken in a representation π defined by a translationally invariant ground state Ψ_0, i.e. for some index i and for some (self-adjoint) $A \in \mathcal{A}_0$

$$J^i(t) \equiv i \lim_{R \to \infty} < [Q_R^i(t), A] >_0 \neq 0. \tag{18.5}$$

Let $\tilde{c}$ be the "reduced" matrix, with matrix elements $\tilde{c}_{jk} = 0$ if $J^j(t)$ and/or $J^k(t)$ is identically zero for all t and $\tilde{c}_{jk} = c_k^j$, (defined in (18.3)), otherwise.
Then, there are quasi particle excitations with infinite lifetime in the limit $\mathbf{k} \to 0$ (generalized Goldstone quasi particles) with an energy spectrum at $\mathbf{k} \to 0$ given by the positive eigenvalues of $\tilde{c}$ which belong to the spectral support of $J^i(0)$.

Proof. By using (18.3) and (18.4), we have

$$i \frac{d}{dt} J^i(t) = \lim_{R \to \infty} \lim_{L \to \infty} < [[Q_R^i(t), H_L], A] >_0 = \lim_{R \to \infty} c_k^i < [Q_R^k(t), A] >_0$$

$$= \tilde{c}_{ik} J^k(t).$$

---

[186] G. Morchio and F. Strocchi, Ann. Phys. **185**, 241 (1988).

The solution of the above equation is

$$J(t) = \exp\left[-i\,\tilde{c}\,t\right] J(0).$$

By the integrability condition of the charge commutators, $J^i(t)$ is polynomially bounded in t and therefore the spectral support of $J(0)$ must consist of real points.

By writing $\tilde{c}$ in Jordan form, one gets

$$J^i(t) = \sum_{\alpha=1}^{k} \mathcal{P}_\alpha^i(t)\, e^{-i\omega_\alpha t},$$

where $\mathcal{P}_\alpha^i(t)$ are polynomials and ω_α belong to the spectral support of $J^i(0) = \sum_\alpha \mathcal{P}_\alpha^i(0)$ relative to $\tilde{c}$. By definition of spectral support, for each α, the zero order coefficient $\mathcal{P}_\alpha^i(0)$ is different from zero, for at least one index i. Thus, for each α there exists at least one index i such that $\tilde{J}^i(\omega)$ contains a contribution of the form $\mathcal{P}_\alpha^i(0)\,\delta(\omega - \omega_\alpha)$.[187]

By (15.31), which relates $\tilde{J}^i(\omega)$ to the energy spectrum at $\mathbf{k} \to 0$, it follows that there are discrete quasi particle excitations with infinite lifetime and energy ω_α, in the limit $\mathbf{k} \to 0$. Each contribution can be isolated by taking suitable linear combinations of the Q_R^i.

The above theorem provides exact information on how the energy spectrum of the Goldstone quasi particles gets modified by the addition of a symmetry breaking interaction (typically with an external field) with simple transformation properties, in the sense of (18.3). Since the symmetric part of the Hamiltonian does not enter in (18.3), the modification of the energy spectrum, typically the energy gap, does not depend on it.

18.1 Example. Spin Model with Magnetic Field

As a concrete example we consider[188] a Heisenberg-like spin model in the presence of a magnetic field h (for simplicity taken in the 3-direction), with the following (finite volume of size L) Hamiltonian

$$H_L = H_{inv,L}(\mathbf{s}) + h \sum_{|i| \leq L} s_i^3, \tag{18.6}$$

where $H_{inv,L}(\mathbf{s})$ is a rotationally invariant spin Hamiltonian with finite range interactions, having a translationally invariant ground state.

---

[187] The possible additional terms $\delta^{(n)}(\omega - \omega_\alpha)$ do not add any further information, since they only give a more singular description of the same spectrum; in fact, such contributions can be isolated by constructing new charges by time derivatives of the original $Q_R^i(t)$.

[188] G. Morchio and F. Strocchi, Ann. Phys. **185**, 241 (1988).

The rotations and the dynamics generate a Lie group G, as the covariance group of the dynamics. As a consequence of the finite range, the time evolution induces a delocalization of fast decrease;[189] then the commutators of

$$S_R^a(t) \equiv \sum_{|i| \leq R} s_i^a(t), \quad \alpha = 1, 2, 3,$$

with a local A are absolutely summable in norm (as distributions in t) and the same property holds for the algebra $\mathcal{A}_0$ generated by the time evolved of elements of $\mathcal{A}_L$. Under general technical conditions one can also prove that (18.3) hold on $\mathcal{A}_0$.[190]

The presence of the external magnetic field implies the breaking of the symmetries generated by S_R^1, S_R^2 and the matrix $\tilde{c}$ is given by

$$\tilde{c}_{ii} = 0, \quad i = 1, 2, \quad \tilde{c}_{12} = -ih = \tilde{c}_{21}.$$

Then, Theorem 18.1 implies that there are Goldstone quasi particles with energy $\omega(\mathbf{k})$ satisfying

$$\lim_{\mathbf{k} \to 0} \omega(\mathbf{k}) = h.$$

---

[189] By (7.16), if $A \in \mathcal{A}(V_0)$, there are suitable positive constants C, v such that for $|t| < v^{-1}|\mathbf{x}|$,

$$||[\, s_i^a, \, \alpha_t(A)\,]|| = ||[\alpha_{-t}(s_i^a), \, A]|| \leq Ce^{-\text{dist}(i, V_A)/2}, \tag{18.7}$$

where $\text{dist}(i, V_A)$ is the distance between the lattice point i and the localization region V_A of A. This implies that a fast decrease of the delocalization induced by the dynamics holds for all A of the form $A = \alpha_\tau(B)$, $\tau \in \mathbf{R}$, $B \in \mathcal{A}_L$ and therefore for the algebra $\mathcal{A}_0$ generated by them.

[190] G. Morchio and F. Strocchi, Ann. Phys. **185**, 241 (1988).

19 Symmetry Breaking in Gauge Theories

The standard model of elementary particle physics crucially relies on the description of elementary particle interactions by gauge field theories and the unification of electromagnetic and weak interactions is made possible by the mechanism of spontaneous symmetry breaking. The intriguing fact that such a breaking of a continuous symmetry (the $SU(2) \times U(1)$ group) is not accompanied by massless Goldstone bosons, in apparent contrast with the conclusions of the Goldstone theorem, demands a general (possibly non-perturbative) understanding and control of such a phenomenon, the so-called *Higgs mechanism*.

19.1 Higgs Mechanism. Problems of the Perturbative Approach

The standard discussion of this mechanism is based on the perturbative expansion and, in particular, the evasion of the Goldstone theorem is checked at the tree level with the disappearance of the massless Goldstone bosons and with the vector bosons becoming massive.[191] This is clearly displayed by the Higgs-Kibble (abelian) model of a (complex) scalar field φ interacting with a real gauge field A_μ, defined by the following Lagrangean $(\rho(x) \equiv |\varphi(x)|)$

$$\mathcal{L} = -\tfrac{1}{4}F_{\mu\nu}{}^2 + \tfrac{1}{2}|D_\mu\varphi|^2 - U(\rho) \quad D_\mu = \partial_\mu - ieA_\mu. \tag{19.1}$$

$\mathcal{L}$ is invariant under the global gauge group $U(1)$: $\beta^\lambda(\varphi) = e^{i\lambda}\varphi$, $\beta^\lambda(A_\mu) = A_\mu$ and under local gauge transformations.

At the classical level, one may argue that by a local gauge transformation

$$\varphi(x) = e^{i\theta(x)}\rho(x) \to \rho(x), \quad A_\mu(x) \to A_\mu(x) + e^{-1}\partial_\mu\theta(x) \equiv W_\mu(x)$$

---

[191] P.W. Higgs, Phys. Lett. **12**, 132 (1964); Phys. Rev. Lett. **13**, 508 (1964); Phys. Rev. **145**, 1156 (1966); Spontaneous Symmetry Breaking, in *Phenomenology of particle physics at high energy: Proc. 14th Scottish Univ. Summer School in Physics 1973*, R.L. Crawford and R. Jennings eds., Academic Press 1974, p. 529; G.S. Guralnik, C.R. Hagen and T.W. Kibble, Phys. Rev. Lett. **13**, 585 (1964); Broken Symmetries and the Goldstone Theorem, in *Advances in Particle Physics*, Vol. 2, R.L. Cool and R.E. Marshak eds., Interscience 1968, p. 567; T.W. Kibble, Broken Symmetries, in *Proc. Oxford Int. Conf. on Elementary Particles*, 1965, Oxford Univ. Press 1966, p. 19; Phys. Rev. **155**, 1554 (1966); F. Englert and R. Brout, Phys. Rev. Lett. **13**, 321 (1964).

F. Strocchi: *Symmetry Breaking in Gauge Theories*, Lect. Notes Phys. **732**, 193–206 (2008)
DOI 10.1007/978-3-540-73593-9_19   © Springer-Verlag Berlin Heidelberg 2008

one may eliminate the field θ from the Lagrangean, which becomes

$$\mathcal{L} = -\tfrac{1}{4}F_{\mu\nu}{}^2 + \tfrac{1}{2}e^2\rho^2\,W_\mu^2 + \tfrac{1}{2}(\partial_\mu\rho)^2 - U(\rho). \tag{19.2}$$

If the (classical) potential U has a non-trivial (absolute) minimum $\rho = \overline{\rho}$ one can consider a semiclassical approximation based on the expansion $\rho = \overline{\rho} + \sigma$, treating $\overline{\rho}$ as a classical constant field and σ as small. At the lowest order, keeping only the quadratic terms in σ and W_μ one has

$$\mathcal{L}^{(2)} = -\tfrac{1}{4}F_{\mu\nu}{}^2 + \tfrac{1}{2}e^2\overline{\rho}^2\,W_\mu^2 + \tfrac{1}{2}(\partial_\mu\sigma)^2 - \tfrac{1}{2}U''(\overline{\rho})\sigma^2. \tag{19.3}$$

This Lagrangean describes a massive vector boson and a massive scalar with (square) masses

$$M_W^2 = \tfrac{1}{2}e^2\,\overline{\rho}^2, \quad m_\sigma^2 = U''(\overline{\rho}).$$

This argument is taken as an evidence that there are no massless particles in the theory described by the Lagrangean $\mathcal{L}$.

This argument, widely used in the literature,[192] is not without problems, because already at the classical level, for the equivalence between the two forms of the Lagrangean, (19.1), (19.2), one must add the constraint that ρ is positive, which is problematic to reconcile with the time evolution defined by the non linear equations obtained from the Lagrangean (19.2) treating ρ and W_μ as Lagrangean variables. For the variables of the quadratic Lagrangean (19.3), one should also require that the time evolution of σ keeps it bounded by $\overline{\rho}$, a condition which is difficult to satisfy. Thus, the constrained system is rather singular and its mathematical control is doubtful. The situation becomes obviously more critical for the quantum version, since the definition of $|\varphi(x)|$ is very problematic also for distributional reasons. In conclusion, ρ is a very singular field and one cannot consider it as a genuine Lagrangean (field) variable.

A better alternative is to decompose the field $\varphi = \varphi_1 + i\,\varphi_2$ in terms of hermitian fields, and to consider the semiclassical expansion $\varphi_1 = \overline{\varphi} + \chi_1$, $\varphi_2 = \chi_2$, treating χ_i, $i = 1, 2$, as small. By introducing the field $W_\mu \equiv A_\mu + e^{-1}\partial_\mu\chi_2$, one eliminates χ_2 from the quadratic part of the so expanded Lagrangean, which gets exactly the same form of (19.3), with $\overline{\rho}$ replaced by $\overline{\varphi}$ and σ by χ_1.

If indeed the fields χ_i can be treated as small, by appealing to the perturbative (loop) expansion one has that $< \varphi > \sim \overline{\rho} \neq 0$, i.e. the vacuum expectation of φ is not invariant the $U(1)$ global group (*symmetry breaking*). Thus, the expansion can be seen as an expansion around a (symmetry breaking) mean field ansatz, and it is very important that a renormalized perturbation theory based on it exists and yields a non vanishing symmetry breaking order parameter $< \varphi > \neq 0$ at all orders. This is the standard (perturbative) analysis of the Higgs mechanism.

---

[192] See e.g. S. Coleman, *Aspects of symmetry. Selected Erice lectures*, Cambridge Univ. Press 1985, Sect. 2.4.

The extraordinary success of the standard model motivates an examination of the Higgs mechanism from a general non-perturbative point of view. In this perspective, one of the problems is that mean field expansions may yield misleading results about the occurrence of symmetry breaking and the energy spectrum (see Chapters 10, 11 above).

As a matter of fact, a non-perturbative analysis of the possible existence of a symmetry breaking order parameter, by using the euclidean functional integral approach defined by the Lagrangean (19.1), gives symmetric correlation functions and in particular $< \varphi >= 0$ (*Elithur-De Angelis-De Falco-Guerra (EDDG) theorem*). [193] This means that the mean field ansatz is incompatible with the non-perturbative quantum effects and the approximation leading to (19.3) is not correct.

The same negative conclusion would be reached if, (as an alternative to the transformation which leads to (19.2)), by means of a gauge transformation one reduces $\varphi(x)$ to a real, not necessarily positive, field $\varphi_r(x)$. This means that the local gauge invariance has not been completely eliminated and the corresponding Lagrangean, of the same form (19.2) with ρ replaced by φ_r, is invariant under a residual Z_2 local gauge group. Then, an easy adaptation of the proof of the EDDG theorem gives $< \varphi >= 0$ and no symmetry breaking.

In order to avoid the vanishing of a symmetry breaking order parameter, one must reconsider the problem by adding to the Lagrangean (19.1) a gauge fixing $\mathcal{L}_{GF}$ which breaks local gauge invariance. Then, the discussion of the Higgs mechanism necessarily becomes gauge fixing dependent; this should not appear strange, since the vacuum expectation of φ is a gauge dependent quantity.[194]

A non-perturbative analysis of the Higgs mechanism shall be discussed in the following subsections in the prototypic cases of local gauges and of the physical Coulomb gauge; in the latter case we shall get a complete characterization of the Higgs mechanism, namely both the absence of Goldstone bosons and the related absence of massless vector bosons.

---

[193] S. Elitzur, Phys. Rev. **D 12**, 3978 (1975); G.F. De Angelis, D. De Falco and F. Guerra, Phys. Rev. **D 17**, 1624 (1978). The crux of the argument is that gauge invariance decouples the transformations of the fields inside a volume V (in a euclidean functional integral approach) from the transformation of the boundary, so that the boundary conditions are ineffective and cannot trigger non symmetric correlation functions. For a simple account of the argument, see e.g. F. Strocchi, *Elements of Quantum Mechanics of Infinite Systems*, World Scientific 1985, Part C, Sect. 2.5.

[194] The above problem of non-perturbative consistency arises also for gauge fixings involving a mean field ansatz, as for the case of the unitary gauge; see S. Weinberg, *The Quantum theory of Fields*, Vol. II, Sect. 21.1.

19.2 Higgs Mechanism in Local gauges

The evasion of the Goldstone theorem by the Higgs mechanism can be understood by a non-perturbative argument in local (renormalizable) gauges.

For concreteness, we discuss the abelian Higgs-Kibble model in the so-called α gauges obtained by the addition of the gauge fixing $-\frac{1}{2}\alpha(\partial_\mu A^\mu)^2$ to the gauge invariant Lagrangean (19.1). Proceeding as before with a perturbative expansion based on the mean field ansatz $\varphi = \overline{\varphi} + \chi_1 + i\chi_2$, and performing the change of variables $W_\mu = A_\mu + e^{-1}\partial_\mu\chi_2$, one gets a quadratic Lagrangean of the form (19.3) plus the gauge fixing term $-\frac{1}{2}\alpha(\Box\chi_2)^2$.

Thus, χ_2 does not disappear from the quadratic Lagrangean and satisfies a "massless" field equation $\Box^2\chi_2 = 0$; this means that there are massless modes.

The problem is their physical interpretation; an indication against their physical relevance is the α dependence of the corresponding Lagrangean term. Moreover, the general solution of the equation $\Box^2\chi_2 = 0$ is a massless field ($\sim \delta(k^2)$) plus a dipole field ($\sim \delta'(k^2)$). A $\delta'(k^2)$ singularity is not a measure and therefore is not allowed to appear in the physical spectrum because the space-time translations must be described by unitary operator, when restricted to the space of physical vectors where positivity holds, so that their spectral representation on physical states is given by a measure.[195]

Actually, by exploiting the Gauss law relation $j_\mu = \partial^\nu F_{\mu\nu}$, one can find a general *non-perturbative* argument[196] about the unphysical nature of the massless modes associated to the breaking of the $U(1)$ gauge group in local (renormalizable) gauges.

As a first step, one remarks that in local renormalizable gauges, like the Feynman gauge, the field algebra $\mathcal{F}$ is generated by the local charged fields $\varphi(x)$ and by the local vector potential $A_\mu(x)$, the four components of which are quantized as independent fields. Locality of the field algebra together with the relativistic spectral support of the Fourier transforms of the vacuum expectations are the basic properties shared by such local gauges, so that most of the standard wisdom on quantum field theory is available; these are in fact the gauges used in perturbation theory.

Thanks to the locality of the field algebra $\mathcal{F}$, there is no problem for the existence of $\lim_{R\to\infty}[j_0(f_R,\alpha), F]$, $\forall F \in \mathcal{F}$. Furthermore, by locality the limit is independent of the smearing test function α, satisfying the normalization condition $\tilde{\alpha}(0) = 1$, i.e. the commutators of $[j_0(f_R,t), F]$ are independent of t, in the limit $R \to \infty$.

Hence, the $U(1)$ global gauge group is locally generated by the conserved current j_μ, the assumption of the Goldstone theorem are fulfilled and one has to discuss its physical consequences.

---

[195] For the discussion of the physical interpretation of the fields of the quadratic Lagrangean, see T.W. Kibble, Phys. Rev. **155**, 1554 (1966).

[196] F. Strocchi, Comm. Math. Phys. **56**, 57 (1977).

For this purpose, it is important to remark that the price to pay for locality is that one has more degrees of freedom than the physical ones (e.g. the "longitudinal photons") and the Maxwell equations hold in a weak form (*weak Gauss' law*),

$$j_\mu(x) = \partial^\nu F_{\mu\nu}(x) + \mathcal{L}_\mu(x), \qquad (19.4)$$

where $\mathcal{L}_\mu(x)$ is an "unphysical" field which must have vanishing matrix elements $< \Psi, \mathcal{L}_\mu\Phi >$ between physical states, in order to avoid violation of Maxwell equations in physical expectations (see (19.5) below).

E.g., in the Feynman-Gupta-Bleuler (FGB) gauge one has

$$-\Box A_\mu(x) = j_\mu(x) = \partial^\nu F_{\mu\nu}(x) + \partial_\mu\partial^\nu A_\nu(x), \qquad (19.5)$$

and the subspace of physical vectors Ψ is identified by the subsidiary (Gupta-Bleuler) condition

$$(\partial^\nu A_\nu)^- \Psi = 0,$$

where ∂A^- denotes the negative energy part of the free field ∂A. Such a condition implies the vanishing of the expectations $< \Psi, \mathcal{L}_\mu\Psi >$.

Such features are clearly displayed by the local (covariant) quantization of the free vector potential[197] but can be argued to be present in general if locality holds.[198]

After these premises we can state

Theorem 19.1. *(Higgs theorem in local gauges) In the local gauges the breaking of the $U(1)$ group, with order parameter*

$$< \delta A >= i \lim_{R\to\infty} < [\, j_0(f_R\alpha), A\,] >\neq 0, \quad A \in \mathcal{F},$$

*implies that the Fourier transform of the two point function $< j_0(x)\, A >$ contains a $\delta(k^2)$ (**Goldstone modes**).*

*However, such a singularity cannot be ascribed to the energy-momentum spectrum of the physical vectors Ψ, which satisfy the weak Gauss law $< \Psi, (j_\mu - \partial^\nu F_{\mu\nu})\Psi >= 0$, i.e. the Goldstone modes are **not physical**.*

Proof. The proof crucially exploits locality and weak Gauss law. The existence of the $\delta(k^2)$ singularity follows from a slight extension of the proof of the Goldstone theorem in the absence of positivity (see Chapter 17).

Moreover, by the locality of A one has

$$\lim_{R\to\infty} < [\partial^i F_{0i}(f_R\alpha), A\,] >= 0$$

---

[197] See e.g. S.S. Schweber, *An Introduction to Relativistic Quantum Field Theory*, Harper and Row 1961, Chap. 9.

[198] F. Strocchi, *Selected Topics on the General Properties of Quantum Field theory*, World Scientific 1993, Chaps. VI, VII; the interplay between locality and Gauss' law is discussed, e.g., in F. Strocchi, *Elements of Quantum Mechanics of Infinite Systems*, World Scientific 1985, Part C, Chap. II.

and therefore the symmetry breaking condition may also be written as

$$< \delta A >= \lim_{R \to \infty} < [\mathcal{L}_R, A] >\neq 0, \quad \mathcal{L}_R \equiv (j_0 - \partial^i F_{0\,i})(f_R\,\alpha). \quad (19.6)$$

For the implication of the $\delta(k^2)$ on the energy-momentum spectrum of the physical vectors, since the vacuum expectation of the commutator is proportional to Im $< j_0(f_R\alpha)A >$, as in the standard proof one has to insert there a complete set of vectors Φ_n (or better discuss a possible spectral representation of the space time translations).

This requires some care since, by the same argument of the GNS representation, the vacuum expectations of the local field algebra $\mathcal{F}$ define a representation of the field algebra in a vector space $V = \mathcal{F}\Psi_0$, with Ψ_0 the vacuum vector, but the inner product $<, >$ defined by such expectations cannot be semidefinite and therefore V does not have a pre-Hilbert space structure. However, under general conditions[199] one can embed V into a Hilbert space $\mathcal{K}$, with scalar product $(,)$, such that $\forall A, B \in \mathcal{F}$

$$< \Psi_0, A^* B \Psi_0 >=< A\Psi_0, B\Psi_0 >= (A\Psi_0, \eta\, B\Psi_0),$$

where η is the metric operator, satisfying $\eta^* = \eta$, $\eta^2 = 1$, $\eta\Psi_0 = \Psi_0$.[200] Actually, the conclusions of the Theorem are independent of the specific properties of such an embedding; the only relevant property is that, in any case, the subspace $\mathcal{K}_{phys} \subset \mathcal{K}$ of physical states must satisfy the subsidiary condition

$$< \Psi, \mathcal{L}_\mu(x)\Phi >= (\Psi, \eta\, \mathcal{L}_\mu(x)\Phi) = 0, \quad \forall \Psi, \Phi \in \mathcal{K}_{phys}, \quad (19.7)$$

in order to ensure the validity of the Maxwell equations.

Then, for the proof of the last statement of the Theorem, it is convenient to choose the complete set of intermediate states Φ_n according to the (orthogonal) decomposition of $\mathcal{K} = K_{phys} \oplus K_{phys}^\perp$. Hence, the generic insertion takes the following form

$$(\Psi_0, \eta\mathcal{L}_R\Phi_n)(\Phi_n, A\Psi_0) =< \Psi_0, \mathcal{L}_R\Phi_n > (\Phi_n, A\Psi_0)$$

and by the weak Gauss law the physical vectors cannot contribute. Thus, the Goldstone modes associated to the $\delta(k^2)$ singularity, which appears in the Fourier transform of the two point function $< j_0(x)\, A >$, cannot be ascribed to physical states.

---

[199] G. Morchio and F. Strocchi, Ann. H. Poincareé, **A 33**, 251 (1980); F. Strocchi, *Selected Topics on the General Properties of Quantum Field Theory*, World Scientific 1993, Chap. VI.

[200] In the Feynman (Gupta-Bleuler) gauge of free QED, $\eta = (-1)^{N_0}$, $N_0 = \int d^3k\, a_0^*(k)\, a_0(k)$ (the number of "timelike photons").

19.3 Higgs Mechanism in the Coulomb Gauge

The evasion of the Goldstone theorem in the case of breaking of the $U(1)$ global gauge symmetry can be understood on the basis of the discussion of Chapter 15, as a consequence of the delocalization induced on the charged fields, and in particular on the Higgs field order parameter, by the instantaneous Coulomb interaction term of the Coulomb gauge Hamiltonian.

The Coulomb gauge can be obtained by adding the gauge fixing condition

$$\partial_i A^i(x) = 0, \tag{19.8}$$

equivalently by adding a Lagrangean multiplier $\mathcal{L} \to \mathcal{L} + \xi(\partial_i A^i) \equiv \mathcal{L}_C$: the variation with respect to ξ gives (19.8). Proceeding as before, one gets the following quadratic Lagrangean in the Coulomb gauge

$$\mathcal{L}_C = -\tfrac{1}{4}F_{\mu\nu}{}^2 + \tfrac{1}{2}e^2\overline{\varphi}^2 W_\mu^2 + \tfrac{1}{2}(\partial_\mu \chi_1)^2 - \tfrac{1}{2}U''(\overline{\varphi})\chi_1^2 +$$

$$+\xi(\partial_i W^i - e^{-1}\Delta \chi_2), \tag{19.9}$$

and (19.8) becomes

$$e\partial_i W^i - \Delta \chi_2 = 0.$$

This is a non-dynamical equation and is easily solved by

$$\chi_2(x) = e\left[(\Delta)^{-1}\partial_i W^i\right](x). \tag{19.10}$$

This implies that, whereas χ_1 and W_μ are expected to be local fields, since they are necessarily so in the quadratic approximation given by (19.9), χ_2 cannot be local, since it is a Coulomb delocalized functional of $\partial_i W^i$. Thus, $\varphi(x) = \overline{\varphi}+\chi_1+i\chi_2$ is non-local with respect to W_i, and therefore with respect to $F_{\mu\nu}$; this reflects the general conflict between Gauss law and locality for charged fields.

Since the gauge fixing breaks local gauge invariance, but not the invariance under the global group transformations, the EDDG theorem does not apply and one may consider the possibility of a symmetry breaking order parameter $< \varphi > \neq 0$.

Now, another conceptual problem arises: the starting Lagrangean $\mathcal{L}$ is invariant under the $U(1)$ global group and its breaking with a mass gap seems incompatible with the Goldstone theorem. As an explanation of such an apparent conflict, one finds in the literature the statement that the Goldstone theorem does not apply if the two point function $< j_0(x)\,\varphi(y) >$ is not Lorentz covariant as it happens in the physical gauges, like the Coulomb gauge. As a matter of fact, the Goldstone-Salam-Weinberg proof of the Goldstone theorem crucially uses Lorentz covariance; however, the more general proof discussed in Chapter 17 does not assume it, so that the quest of a better explanation remains.

As discussed in Chapter 16, the condition that the symmetry is generated on the Coulomb field algebra $\mathcal{F}_C$ by the integral of the charge density $Q_R = j_0(f_R\alpha)$ requires that *both* the limit

$$\lim_{R\to\infty} [j_0(f_R,\alpha),F] \tag{19.11}$$

exists $\forall F \in \mathcal{F}_C$ *and it is independent of the time smearing*, i.e. of the test function α, $\tilde{\alpha}(0) = 1$. The latter property is the proper way of stating that the commutator $\lim_{R\to\infty} <[j_0(f_R,t),F]>$ is independent of time.

Quite generally, for the conclusion of the Goldstone theorem one needs a fall off of the current commutators faster than $|\mathbf{x}|^{-2}$, [201]

$$\lim_{|\mathbf{x}|\to\infty} |\mathbf{x}|^2 [\mathbf{j}(\mathbf{x},t),A] = 0. \tag{19.12}$$

For the Higgs-Kibble model in the Coulomb gauge an indication of the failure of (19.12) with $A = \varphi(f)$ can be inferred, in the approximation of (19.9), from the Coulomb delocalization of χ_2 given by (19.10) and therefore of $\varphi(x) = \overline{\varphi} + \chi_1 + i\chi_2$.

Actually a full *non-perturbative characterization of the Higgs mechanism*, namely *both* the absence of massless Goldstone bosons *and* the absence of massless vector bosons can be obtained in the physical Coulomb gauge, [202] by exploiting the relation between the non-local charged field φ in the Coulomb gauge and the local charged field ψ in the Feynman-Gupta-Bleuler (FGB) gauge, formally given by[203]

$$\varphi(x) = e^{-i e\,[(-\Delta)^{-1}\partial_i A^i](x)}\,\psi(x), \tag{19.13}$$

where A^i, ψ are the (renormalized) fields which describe the vector potential and the charged field, respectively, in the local FGB quantization of QED (and e is the renormalized charge).

For the discussion of the relation between the generator Q of the $U(1)$ global gauge group and a suitable integral of the charge density of the associated conserved Noether current $j_\mu = \partial^\nu F_{\mu\nu}$, it is enough to consider the commutators with the Coulomb charged field φ (and the vector potential) which generate the Coulomb field algebra $\mathcal{F}_C$.

---

[201] This point was first pointed out by G.S. Guralnik, C.R. Hagen and T.W. Kibble, Phys. Rev. Lett. **13**, 585 (1964); see also T.W. Kibble, Phys. Rev. **155**, 1554 (1966); G.S. Guralnik, C.R. Hagen and T.W. Kibble, Broken Symmetries and the Goldstone theorem, in *Advances in Particle Physics*, Vol. 2, R.L. Cool and R.E. Marshak eds., Interscience 1968, p. 567.

[202] G. Morchio and F. Strocchi, J. Phys. A: Math. Theor. **40**, 3173 (2007).

[203] P.A.M. Dirac, Canad. J. Phys. **33**, 650 (1955); K. Symanzik, Lectures 1971, loc. cit. For the necessary UV regularization and its rigorous version, see O. Steinmann, *Perturbative Quantum Electrodynamics and Axiomatic Field Theory*, Springer 2000; D. Buchholz, S. Doplicher, G. Morchio, J.E. Roberts and F. Strocchi, Ann. Phys. **290**, 53 (2001).

Proposition 19.2. *In the Coulomb gauge* $\forall \Psi$, $\Phi \in \mathcal{F}_C \Psi_0$, Ψ_0 *denoting the vacuum vector, the limits*

$$\lim_{R \to \infty} (\Psi, [j_0(f_R \alpha), \varphi] \Phi) \tag{19.14}$$

exist but are α dependent and therefore the time independent $U(1)$ global gauge group cannot be generated by such integrals of the charge density.

Proof. The convergence and time dependence of the charge density commutators

$$[j_0(f_R, x_0), \varphi(y)] = [\partial^i F_{0i}(f_R, x_0), \varphi(y)]$$

are governed by the large (spacelike) distance behaviour of the commutator $[F_{0j}, \varphi(y)]$, i.e. of $[F_{0j}(x), e^{-ie(-\Delta^{-1}\partial_i A^i)(y)}]$, as a consequence of (19.13).

Now, for spacelike separations $|\mathbf{x}| \to \infty$, one has the following general estimate, (obtained by expanding the exponential and by exploiting the cluster property of the correlation functions of the FGB fields),

$$[F_{\mu\nu}(x), \varphi(y)] \sim \frac{ie}{4\pi} \int d^3z\, \partial_z^j \frac{1}{|\mathbf{y} - \mathbf{z}|} < [F_{\mu\nu}(x), A_j(\mathbf{z}, y_0)] > \varphi(y), \tag{19.15}$$

the correction being at least $O(|\mathbf{x}|^{-4})$.[204] This and all the following equations are understood to hold in matrix elements of *Coulomb states* $\Psi, \Phi \in \mathcal{F}_C \Psi_0$.

Since $< [F_{\mu\nu}(x), A_j(z)] > = (\partial_\nu g_{\mu j} - \partial_\mu g_{\nu j}) F(x - z)$, where F is the Lorentz invariant distribution which characterizes the vacuum expectation of the electromagnetic field commutator

$$< [F_{\mu\nu}(\tfrac{1}{2}x), F_{\lambda\sigma}(-\tfrac{1}{2}x)] >_0 = id_{\mu\nu\lambda\sigma} \int d\rho(m^2) \Delta(x; m^2) \equiv d_{\mu\nu\lambda\sigma} F(x), \tag{19.16}$$

---

[204] The point is that, by locality of the Feynman-Gupta-Bleuler fields, the commutator $[F_{\mu\nu}(x + \mathbf{a}), A_j(\mathbf{z}, y_0)]$ has a compact support in $\mathbf{z}$ and therefore, for $|\mathbf{a}| \to \infty$, the convolution with $\partial_z^j \frac{1}{|\mathbf{y}-\mathbf{z}|}$ decreases at least as $|\mathbf{a}|^{-2}$. By the same reasons,

$$\int d^3z\, \partial_z^j \frac{1}{|\mathbf{y} - \mathbf{z}|} [F_{\mu\nu}(x + \mathbf{a}), A_j(\mathbf{z}, y_0)]$$

commutes with the other factors $(-\Delta^{-1}\partial^j A_j)(y)$, in the expansion of the exponential, apart from terms decreasing at least as $|\mathbf{a}|^{-4}$. Thus, for $|\mathbf{a}| \to \infty$)

$$[F_{0j}(x + \mathbf{a}), e^{-ie(-\Delta^{-1}\partial_i A^i)(y)}] \sim$$

$$\sim \int d^3z\, \partial_z^j \frac{1}{|\mathbf{y} - \mathbf{z}|} [F_{\mu\nu}(x + \mathbf{a}), A_j(\mathbf{z}, y_0)] e^{-ie(-\Delta^{-1}\partial_i A^i)(y)},$$

and, by the cluster property of the FGB correlation functions, the vacuum insertion gives the leading contribution. For a proof of this behaviour, which takes into account the need of an UV regularization of (19.13) and exploits the locality of the charged fields in the FGB gauge and the cluster property see D. Buchholz, S. Doplicher, G. Morchio, J.E. Roberts and F. Strocchi, Ann. Phys. **290**, 53 (2001).

$d_{\mu\nu\lambda\sigma} \equiv g_{\nu\lambda}\partial_\mu\partial_\sigma + g_{\mu\sigma}\partial_\nu\partial_\lambda - g_{\nu\sigma}\partial_\mu\partial_\lambda - g_{\mu\lambda}\partial_\nu\partial_\sigma$, one has, for $R \to \infty$,

$$[j_0(f_R, x_0), \varphi(y)] = [\partial^i F_{0i}(f_R, x_0), \varphi(y)] \sim$$

$$\sim -ie\partial_0 \int d^3x\, f_R(\mathbf{x})F(x - y)\varphi(y). \tag{19.17}$$

By the support properties of $F(x) = \int d\rho(m^2)\, \epsilon(k_0)\, \delta(k^2 - m^2)\, e^{-ikx}$, the charge density is integrable and, in all correlation functions of the Coulomb field algebra,

$$\lim_{R\to\infty} [j_0(f_R, x_0), \varphi(y)] = -e \int d\rho(m^2)\, \cos(m(x_0 - y_0))\, \varphi(y). \tag{19.18}$$

The r.h.s is independent of time if and only if $d\rho(m^2) = \lambda\delta(m^2)$, i.e. if $F_{\mu\nu}$ is a free field. [205]

The same conclusions hold if instead of (19.13) one uses the regularized version of Buchholz et al., since (19.15), (19.17), (19.18) get changed only by a convolution with a test function $h(y_0) \in \mathcal{D}(\mathbf{R})$.

The same conclusion about the time dependence of the commutators of the charge density is obtained by the general estimate of the spacelike large distance behaviour of the commutator (19.12) obtained by using (19.15), i.e.

$$[j_i(x), \varphi(y)] \sim i(e/4\pi) \int d^3z\, \partial^i_z|\mathbf{z} - \mathbf{y}|^{-1}\partial^2_0 F(\mathbf{x} - \mathbf{z}, x_0 - y_0)\, \varphi(y),$$

$$\lim_{R\to\infty} [\dot{Q}_R(x_0), \varphi(y)] = [\operatorname{div}\mathbf{j}(f_R, x_0), \varphi(y)] \neq 0.$$

The time dependence of the charge density commutators is at the roots of the appearance of an infinite renormalization constant in equal time commutators

$$[j_0(\mathbf{x}, t), \varphi(\mathbf{y}, t)] = -e(Z_3)^{-1}\delta(\mathbf{x} - \mathbf{y})\varphi(\mathbf{y}, t).$$

For such a phenomenon the vacuum polarization due to loops of charged fields plays a crucial role, so that the semi-classical approximation does not provide relevant information about the time dependence of the charge commutators. In fact, the phenomenon does not appear in the classical theory, where there are finite energy localized solutions with non zero charge and localized current j_μ, only the electric field being a Coulomb delocalized function of j_0. [206]

The above Proposition shows that the heuristic argument by which if the symmetry commutes with time translation, equivalently if the current continuity equation holds, then the generating charge commutes with the Hamiltonian and is therefore independent of time is not correct.

---

[205] G. Morchio and F. Strocchi, J. Math,. Phys. **44**, 5569 (2003), Appendix.
[206] D. Buchholz, S. Doplicher, G. Morchio, J.E. Roberts and F. Strocchi, Ann. Phys. **290**, 53 (2001).

Even if the equal time commutators, in particular $[\mathbf{j}, \varphi]$, have a sufficient localization, the time evolution may induce a delocalization leading to a failure of (19.12). For these reasons, no reliable information can be inferred from the equal time commutators and the check of the basic assumptions of the Goldstone theorem becomes interlaced with the dynamical problem, as it happens for non-relativistic systems.

The failure of locality, rather than the lack of manifest covariance, is the crucial structural property which explains the evasion of the Goldstone theorem by exactly the same mechanism in the Higgs mechanism as well as in non-relativistic Coulomb systems and in the $U(1)$ problem (see the discussion of Chapter 15 and below).

Theorem 19.3. *(***Higgs phenomenon***) If the spectral measure of the vector boson field $F_{\mu\nu}$ has a $\delta(k^2)$ contribution, i.e. if there are corresponding* **massless vector bosons***, then the global $U(1)$* **symmetry** *is* **unbroken***.*

If the (time independent) $U(1)$ global gauge symmetry is broken, i.e. there is a field F of the Coulomb field algebra $\mathcal{F}_C$ such that

$$< \delta F >= i < [\, Q, F\,] >\neq 0, \qquad (19.19)$$

where Q is the generator of $U(1)$, then
i) the Fourier transform of the two point function of the vector boson field, $< F_{\mu\nu}(\tfrac{1}{2}x)\, F_{\rho\sigma}(-\tfrac{1}{2}x) >$, cannot contain a $\delta(k^2)$, i.e. there are **no massless vector bosons** *associated with $F_{\mu\nu}$,*
ii) the two point function $< j_\mu(x)F >$, cannot vanish and its Fourier spectrum, i.e. the Goldstone spectrum, coincides with that of the two point function of the vector boson field $F_{\mu\nu}$, so that the absence of massless vector bosons coincides with the **absence of massless Goldstone bosons***.*

Proof. The proof of i) is equivalent to the proof that if the Fourier transform of the two point function of the vector boson field contains a $\delta(k^2)$, i.e. if there are corresponding massless vector bosons, then the $U(1)$ symmetry cannot be broken.

For this purpose we first note that in this case a spacelike time average of the integral of the charge density $Q_{R\delta} \equiv j_0(f_R, \alpha_{T(R)})$ with $\alpha_{T(R)}(x_0) = \alpha(x_0/T(R))/T(R)$, $T(R) = \delta R$, $0 < \delta < 1$, the limit $\delta \to 0$ to be taken after the limit $R \to \infty$, generates the $U(1)$ group on the Coulomb field algebra (in matrix elements of Coulomb states $\Psi, \Phi \in \mathcal{F}_C$): [207]

$$\delta F = i \lim_{\delta \to 0} \lim_{R \to \infty} [\, j_0(f_R \alpha_{\delta R}, F\,], \qquad (19.20)$$

if and only if the spectral measure $d\rho(k^2)$, which characterizes the vacuum expectation of the electromagnetic field commutator (19.16), has a $\delta(k^2)$ contribution, i.e. there are corresponding massless vector bosons.

In fact, the time smearing of (19.18) with $\alpha_{\delta R}(x_0)$ gives

---

[207] G. Morchio and F. Strocchi, J. Phys. A: Math. Phys. **40**, 3173 (2007).

$$[j_0(f_R \alpha_{\delta R}), \varphi_C(y)] =$$

$$= e \int d\rho(m^2) \, d^3q \, \tilde{f}(\mathbf{q}) \, \mathrm{Re}[e^{-i\omega_R(q,m)y_0} \tilde{\alpha}(\delta\sqrt{\mathbf{q}^2 + R^2 m^2})] \, \varphi_C(y),$$

where $\omega_R(\mathbf{q}, m) \equiv \sqrt{\mathbf{q}^2 R^{-2} + m^2}$. Then, since α is of fast decrease, by the dominated convergence theorem the r.h.s. vanishes if the $d\rho(m^2)$ measure of the point $m^2 = 0$ is zero, i.e. if there is no $\delta(m^2)$ contribution to $d\rho$. In general, if the point $m^2 = 0$ has measure λ, one gets $\lambda e \, \varphi_C(y)$; finally, the renormalization condition of the asymptotic electromagnetic field gives $\lambda = 1$.

Moreover, if $d\rho(k^2)$ has a $\delta(k^2)$ contribution one has

$$\mathrm{strong} - \lim_{R \to \infty} j_0(f_R \, \alpha_{\delta R}) \, \Psi_0 = 0. \tag{19.21}$$

In fact, one has $(d\Omega_m(\mathbf{k}) \equiv d^3 k (2\sqrt{\mathbf{k}^2 + m^2})^{-1})$

$$||Q_{R\delta} \Psi_0||^2 = \int d\rho(m^2) m^2 d\Omega_m(\mathbf{k}) \, |\, \mathbf{k} \, \tilde{f}_R(\mathbf{k}) \tilde{\alpha}(\delta R \sqrt{\mathbf{k}^2 + m^2})|^2 =$$

$$= \int d\rho(m^2) \, d\Omega_m(\mathbf{q}/R) \, m^2 \, R \, |\tilde{\alpha}(\delta\sqrt{\mathbf{q}^2 + m^2 R^2}) \, \mathbf{q} \, \tilde{f}(\mathbf{q})|^2.$$

Now, $m^2 R \, |\tilde{\alpha}(\delta\sqrt{\mathbf{q}^2 + m^2 R^2})|^2$ converges pointwise to zero for $R \to \infty$ and, since $d\rho(m^2)$ is tempered and α is of fast decrease, the r.h.s. of the above equation converges to zero by the dominated convergence theorem.

The proof of ii) follows from (19.17), (19.18), since it is enough to consider the case $F = \varphi$. In fact, in the proof of the Goldstone theorem, the role of the local generation of the symmetry by the density of the corresponding Noether current is that of assuring the non vanishing of the two point function $< j_\mu(x) F >$ as a consequence of $< \delta F > \neq 0$. Now, even if the $U(1)$ group is not generated by a suitable integral of the current charge density, nevertheless, thanks to the estimate (19.17), $< \varphi > \neq 0$ implies that the two point function $< j_\mu(x) \varphi >$ cannot vanish, being proportional to the vector boson commutator function (19.16).

Moreover, by (19.17) the Goldstone boson spectrum, i.e. the Fourier spectrum of $< j_\mu(x) \varphi >$ is given by the vector boson spectral measure $d\rho(k^2)$ and the latter cannot contain a $\delta(k^2)$ because, otherwise, by (19.20), (19.21),

$$< \delta\varphi > = i \lim_{\delta \to 0} \lim_{R \to \infty} < [j_0(f_R \alpha_{\delta R}), \varphi] > = 0.$$

In conclusion, the vector bosons associated with $F_{\mu\nu}$ cannot be massless and there are no massless Goldstone bosons.

19.4 Axial Symmetry Breaking and U(1) Problem

The debated problem of $U(1)$ axial symmetry breaking in quantum chromodynamics without massless Goldstone bosons can be clarified by the realization of the non locality of the associated axial current.

As clearly shown by Bardeen,[208] the $U(1)$ axial symmetry gives rise to a conserved, gauge dependent, current

$$J_\mu^5 = j_\mu^5 - (2\pi)^{-2}\varepsilon_{\mu\nu\rho\sigma}\text{Tr}\left[A^\nu\partial^\rho A^\sigma - (2/3)iA^\nu A^\rho A^\sigma\right] \equiv j_\mu^5 + K_\mu^5,$$

where j_μ^5 is the gauge invariant point splitting regularized fermion current $\overline{\psi}\gamma_\mu\gamma_5\psi$. The current j_μ^5 is not conserved because of the anomaly, which is equivalent to the conservation of J_μ^5.

In the usual discussion of the $U(1)$ problem,[209] the current J_μ^5 has been discarded on the blame of its gauge dependence, and the lack of conservation of j_μ^5 has been taken as the evidence that the axial $U(1)$ is not a symmetry of the field algebra and therefore the problem of its spontaneous breaking does no longer exist. Such a conclusion would imply that time independent $U(1)$ axial transformations cannot be defined on the field algebra $\mathcal{F}$ and not even on its observable subalgebra $\mathcal{F}_{obs}$, which contains the relevant order parameter.

However, as argued by Bardeen on the basis of perturbative renormalization (in local gauges), *the axial $U(1)$ transformations define a time independent symmetry of the field algebra and of its observable subalgebra.* This also follows from the conservation of J_μ^5 (equivalent to the anomaly of j_μ^5), since in local renormalizable gauges J_μ^5 is a local operator, so that the standard argument applies. This implies that (at least at the infinitesimal level) the limit

$$\lim_{R\to\infty}\left[J_0^5(f_R,\alpha), F\right], \quad F \in \mathcal{F},$$

defines in this case a symmetry of the field algebra and in particular of the gauge invariant observable subalgebra $\mathcal{F}_{obs}$.

Therefore, there is no logical reason for *a priori* rejecting the use of the gauge dependent current J_μ^5 and of its associated Ward identities; one should only keep in mind that in physical gauges J_μ^5 is a non local function of the observable (gauge independent) fields. The structure is somewhat specular to that of the Higgs case, where the current is a local observable field but the Higgs field giving the order parameter is not local.

The existence of axial $U(1)$ transformations of the observable subalgebra $\mathcal{F}_{obs}$ implies that *the absence of parity doublets is a problem of spontaneous symmetry breaking* and *the absence of massless Goldstone bosons is reduced to the discussion of local generation of the symmetry,* as in the case of the Higgs phenomenon.

In the local (renormalizable) gauges, the time independent $U(1)$ axial symmetry is generated by J_μ^5 (and not by j_μ^5) and the problem of massless Goldstone modes does not arise because, as indicated by the perturbative

---

[208] W.A. Bardeen, Nucl. Phys. **B 75**, 246 (1974).

[209] See e.g. S. Coleman, *Aspects of Symmetry*, Cambridge Univ. Pres 1985, Chap. 7.

expansion and also by the Schwinger model [210] *the correlation functions of the (local) field algebra $\mathcal{F}$ are axial $U(1)$ invariant.*

However, the invariance of the vacuum functional Ψ_0, which defines the local gauge quantization, does not mean that the symmetry is unbroken in the irreducible representation of the observable subalgebra $\mathcal{F}_{obs}$. In fact, Ψ_0 gives a reducible representation of $\mathcal{F}_{obs}$ (as signaled by the failure of the cluster property by the corresponding vacuum expectations), with a non-trivial center which is generated by the large gauge transformations T_n and is not pointwise invariant under $U(1)$ axial transformations.

Thus, *the symmetry is broken in each pure physical phase (θ-vacuum sectors)* obtained by the diagonalization of the T_n (in the technical terminology by a *central decomposition* of the observables) in the subspace $\mathcal{F}_{obs}\Psi_0$.

It should be stressed that the so obtained (gauge invariant) θ-vacua do *not* provide well defined representations of the field algebra $\mathcal{F}$, since the latter transforms non trivially under T_n. This is at the origin of the difficulties (and paradoxes) arising in the discussion of the chiral Ward identities (corresponding to the conservation of J_μ^5) in θ-vacua expectations.[211]

In the θ sectors, a conserved axial current may be constructed as a non local operator, typically by using for J_μ^5 its (non local) expression in terms of the observable fields in a physical gauge. The above discussion, in particular the lack of time independence of the charge density commutators, as a consequence of the failure of relative locality between the current and the order parameter, applies to such non local currents.

The resulting mechanism for the solution of the $U(1)$ problem can be made explicit in the Coulomb gauge. In the Schwinger model, in the Coulomb gauge one has $K_0 = (e/\pi)A_1 = 0$, $K_1 = (e/\pi)A_0$, so that $J_0^5 = j_0^5$ and the (θ-)vacuum expectations of the commutators $[J_0^5(f_R, t), A]$, $[j_0^5(f_R, t), A]$, $A \in \mathcal{F}_{obs}$, coincide and describe the same mass spectrum; however, the time dependence in the limit $R \to \infty$, in the first case can be ascribed to the non locality of the conserved axial current, whereas in the second case it reflects the non conservation of j_μ^5.

---

[210] G. Morchio, D. Pierotti and F. Strocchi, Ann. Phys. **188**, 217 (1988); F. Strocchi, *Selected Topics on the General Properties of Quantum Field Theory*, World Scientific 1993, Sect. 7.4.

[211] R.J. Crewter, in *Field Theoretical methods in Particle Physics*, W. Rühl ed. Reidel (1980), p. 529.

References Part I

R.M.F. Houtappel, H. Van Dam and E.P. Wigner, Rev. Mod. Phys. **37**, 595 (1965)

D.H. Sattinger, Spontaneous Symmetry Breaking: mathematical methods, applications and problems in the physical sciences, in *Applications of Non-Linear Analysis*, H. Amann et al. eds., Pitman 1981

G. B. Whitham, *Linear and Non-Linear Waves*, J. Wiley, New York 1974

R. Rajaraman, Phys. Rep. **21 C**, 227 (1975)

S. Coleman, *Aspects of Symmetry*, Cambridge Univ. Press 1985

M. Reed, *Abstract non-linear wave equation*, Springer, Heidelberg 1976

C. Parenti, F. Strocchi and G. Velo, Phys. Lett. **59B**, 157 (1975)

C. Parenti, F. Strocchi and G. Velo, Ann. Scuola Norm. Sup. (Pisa), III, 443 (1976)

F. Strocchi, Lectures at the Workshop on *Recent Advances in the Theory of Evolution Equations*, ICTP Trieste 1979, published in *Topics in Functional Analysis 1980-81*, Scuola Normale Superiore, Pisa 1982

F. Strocchi, Stability properties of the solutions of non-linear field equations. Hilbert space sectors and electric charge, in *Nonlinear Hyperbolic Equations in Applied Sciences*, Rend. Sem. Mat. Univ. Pol. Torino, Fasc. spec. 1988

W. Strauss, *Nonlinear Wave Equations*, Am. Math. Soc. 1989

C. Parenti, F. Strocchi and G. Velo, Comm. Math. Phys. **53**, 65 (1977)

C. Parenti, F. Strocchi and G. Velo, Phys. Lett. **62B**, 83 (1976)

E. Noether, Nachr. d. Kgl. Ges. d. Wiss. Göttingen (1918), p. 235

H. Goldstein, *Classical Mechanics*, 2nd. ed., Addison-Wesley 1980

E. L. Hill, Rev. Mod. Phys. **23**, 253 (1951)

K. Jörgens, Mat. Zeit. **77**, 291 (1961)

I. Segal, Ann. Math. **78**, 339 (1963)

V. Arnold, *Ordinary Differential Equations*, Springer 1992

G. Sansone and R. Conti, *Non-linear Differential Equations*, Pergamon Press 1964

W. Strauss, Anais Acad. Brasil. Ciencias **42**, 645 (1970)

R.F. Streater and A.S. Wightman, *PCT, Spin and Statistics and All That*, Benjamin-Cumming Pubbl. C. 1980

D. Ruelle, *Statistical Mechanics*, Benjamin 1969

R. Haag, *Local Quantum Physics*, Springer-Verlag 1996

C. Parenti, F. Strocchi and G. Velo, Structure Properties of Solutions of Classical Non-linear Relativistic Field Equations, in *Invariant Wave Equations*, G. Velo and A.S. Wightman eds., Springer-Verlag 1978

L.R. Volevic and B.P. Paneyakh, Russian Math. Surveys **20**, 1 (1965)

S. Coleman, Phys. Rev. D **15**, 2929 (1977)

R. Haag and D. Kastler, J. Math. Phys. **5**, 848 (1964)

J. Goldstone, Nuovo Cim. **19**, 154 (1961)

J. Goldstone, A. Salam and S. Weinberg, Phys. Rev. **127**, 965 (1962)

Y. Nambu and G. Jona-Lasinio, Phys. Rev. **122**, 345 (1961); Phys. Rev. **124**, 246 (1961)

F. Strocchi, *Elements of Quantum Mechanics of Infinite Systems*, World Scientific 1985

P.W. Higgs, Phys. Lett. **12**, 132 (1964)

T.W. Kibble, Broken Symmetries, in *Proc. Oxford Int. Conf. Elementary Particles*, Oxford 1965

G.S. Guralnik, C.R. Hagen and T.W. Kibble, Broken Symmetries and the Goldstone Theorem, in *Advances in Particle Physics* Vol. 2, R.L. Cool and R.E. Marshak eds., Interscience, New York 1968

G. Morchio and F. Strocchi, Infrared problem, Higgs phenomenon and long range interactions, in *Fundamental Problems of Gauge Field Theory*, G. Velo and A.S. Wightman eds., Plenum 1986

F. Strocchi, Mass/energy gap associated to symmetry breaking. A generalized Goldstone theorem for long range interactions, in *Fundamental Aspects of Quantum Theory*, V. Gorini and A. Frigerio eds., Plenum 1986

J. Goldstone and R. Jackiw, Phys. Rev. **D11**, 1486 (1975)

R. Rajaraman, *Solitons and Instantons*, North-Holland 1982

R.F. Dashen, B. Hasslacher and A. Neveu, Phys. Rev. **D10**, 4130 (1974)

J. Goldstone and R. Jackiw, Phys. Rev. **D11**, 1486 (1975)

C. Rebbi and G. Soliani, *Solitons and Particles*, World Scientific 1984

A. Barone, F. Esposito and C.J. Magee, Theory and Applications of the Sine-Gordon Equation, Riv. Nuovo Cim. **1**, 227 (1971)

A.C. Scott, F.Y. Chiu, and D.W. Mclaughlin, Proceedings I.E.E.E. **61**, 1443 (1973)

S. Coleman, Phys. Rev. **D11**, 2088 (1975)

J. Fröhlich, The Quantum Theory of Non-linear Invariant Wave Equations, in *Invariant Wave Equations*, G. Velo and A.S. Wightman eds., Springer-Verlag 1977

J. Goldstone, Nuovo Cimento **19**, 154 (1961)

J. Goldstone, A. Salam and S. Weinberg, Phys. Rev. **127**, 965 (1962)

J. Swieca, Goldstone's theorem and related topics, in *Cargèse Lectures in Physics*, Vol. 4, D. Kastler ed., Gordon and Breach 1970

F. Strocchi, Phys. Lett. **A267**, 40 (2000)

H. Pecher. Math. Zeit. **185**, 261 (1984); **198**, 277 (1988)

References Part II

F. Strocchi, *Elements of Quantum Mechanics of Infinite Systems*, World Scientific 1985

W. Heisenberg, *The Physical Principles of the Quantum Theory*, Dover Publications 1930

P.A.M. Dirac, *The Principles of Quantum Mechanics*, Oxford University Press 1986

M. Reed and B. Simon, *Methods of Modern Mathematical Physics*, Vol. I, Academic Press 1972

J. Slawny, Comm. Math. Phys. **24**, 151 (1972)

J. Manuceau, M. Sirugue, D. Testard and A. Verbeure, Comm. Math. Phys. **32**, 231 (1973)

R. Haag, *Local Quantum Physics*, Springer 1996

F. Strocchi, *An Introduction to the Mathematical Structure of Quantum Mechanics*, (Scuola Normale Superiore, Pisa 1996, World Scientific 2005 [SNS 1996]

O. Bratteli and D.W. Robinson, *Operator Algebras and Quantum Statistical Mechanics*, Vol. 1, 2, Springer 1987, 1996

M.A. Naimark, *Normed Rings*, Noordhoff 1964

G.F. Dell'Antonio, S. Doplicher, J. Math. Phys. **8**, 663 (1967)

J.M. Chaiken, Comm. Math. Phys. **8**, 164 (1967); Ann. Phys. **42**, 23 (1968)

H.J. Borchers, R. Haag and B. Schroer, Nuovo Cim. **29**, 148 (1963)

A.S. Wightman and S. Schweber, Phys. Rev. **98**, 812 (1955)

S.S. Schweber, *Introduction to Relativistic Quantum Field Theory*, Harper and Row 1961

H. Araki and E.J. Woods, J. Math. Phys. **4**, 637 (1963)

H. Araki and W. Wyss, Helv. Phys. Acta **37**, 139 (1964)

R. Jost, *The General Theory of Quantized Fields*, Am. Math. Soc. 1965

R. Haag, *On quantum field theories*, Dan. Mat. Fys. Medd. **29**, No. 12 (1955)

R.F. Streater and A.S. Wightman, *PCT, Spin and Statistics and All That*, Benjamin-Cummings 1980

M. Reed and B. Simon, *Methods of Modern Mathematical Physics*, Vol. II (Fourier Analysis, Self-Adjointness), Academic Press 1975

A.S. Wightman, Introduction to some aspects of the relativistic dynamics of quantized fields, in *Cargèse Lectures in Theoretical Physics*, M. Levy ed., Gordon and Breach 1967

A.S. Wightman, Constructive Field Theory. Introduction to the Problems, in *Fundamental Interactions in Physics and Astrophysics*, G. Iverson et al. eds., Plenum 1972

Constructive Quantum Field Theory, G. Velo and A.S. Wightman eds., Springer 1973

J. Glimm and A. Jaffe, *Quantum Physics*, Springer 1981

R.J. Glauber, Phys. Rev. Lett. **10**, 84 (1963); Phys. Rev. **131**, 2766 (1963)

V. Chung, Phys. Rev. **140B**, 1110 (1965)

J. Fröhlich, G. Morchio and F. Strocchi, Ann. Phys. **119**, 241 (1979)

G. Morchio and F. Strocchi, Nucl. Phys. **B211**, 471 (1984)

G. Morchio and F. Strocchi, Infrared problem, Higgs phenomenon and long range interactions, in *Fundamental Problems of Gauge Field Theory*, G. Velo and A.S. Wightman eds., Plenum 1986

T.W. Kibble, Phys. Rev. **173**, 1527; **174**, 1882; **175**, 1624 (1968)

R. Haag, Subject, Object and Measurement, in *The Physicist's Conception of Nature*, J. Mehra ed., Reidel 1973

A.S. Wightman, Some comments on the quantum theory of measurement, in *Probabilistic Methods in Mathematical Physics*, F. Guerra et al. eds., World Scientific 1992

D.A. Dubin and G.L. Sewell, Jour, Math. Phys. **11**, 2290 (1970)

G.L. Sewell, Comm. Math. Phys. **33**, 43 (1973)

G. Morchio and F. Strocchi, Comm. Math. Phys. **99**, 153 (1985)

G. Morchio and F. Strocchi, J. Math. Phys. **28**, 622 (1987)

D. Kastler, Topics in the algebraic approach to field theory, in *Cargèse Lectures in Theoretical Physics*, F. Lurçat ed., Gordon and Breach 1967

R. Haag, Phys. Rev. **112**, 669 (1958)

D. Ruelle, Helv Phys. Acta **35**, 147 (1962)

H. Araki, Prog. Theor. Phys. **32**, 884 (1964)

D.W. Robinson, Comm. Math. Phys. **7**, 337 (1968)

M. Guenin, Comm. Math. Phys. **1**, 127 (1966)

I.E. Segal, Proc. Natl. Acad. Sci. USA, **57**,1178 (1967)

D.A. Dubin, *Solvable models in algebraic statistical mechanics*, Oxford Univ. Press 1974

N.M. Hugenholtz, Quantum mechanics of infinitely large systems, in *Fundamental Problems in Statistical Mechanics II*, E.G.D. Cohen ed., North-Holland, Amsterdam 1968

R.P. Feynman, *Introduction to Statistical Mechanics*, Benjamin 1972

R.F. Streater, Comm. Math. Phys. **6**, 233 (1967)

E.P. Wigner, *Group Theory and its Applications to the Quantum Mechanics of Atomic Spectra*, Academic Press 1959

V. Bargmann, J. Math. Phys. **5**, 862 (1964)

V. Bargmann, Ann. Math. **59**, 1 (1954)

D.J. Simms, *Lie Groups and Quantum Mechanics*, Lect. Notes Math. 52, Springer 1968

D.J. Simms, Rep. Math. Phys. **2**, 283 (1971)

F. Strocchi, *Selected Topics on the General Properties of Quantum Field Theory*, World Scientific 1993

J. Goldstone, Nuovo Cim. **10**, 154 (1961)

S. Coleman and E. Weinberg, Phys. Rev. **D7**, 1888 (1973)

B.W. Lee, Nucl. Phys. **B9**, 649 (1969)

K. Symanzik, Renormalization of Theories with Broken Symmetry, in *Cargèse Lectures in Physics 1970*, D. Bessis ed., Gordon and Breach 1972

C. Becchi, A. Rouet and R. Stora, Renormalizable Theories with Symmetry Breaking, in *Field Theory, Quantization and Statistical Physics*, E. Tirapegui ed., D. Reidel 1981

J. Collins, *Renormalization*, Cambridge Univ. Press 1984

L.S. Brown, *Quantum Field Theory*, Cambridge Univ. Press 1994

D. Ruelle, *Statistical Mechanics*, Benjamin 1969

G.L. Sewell, *Quantum Theory of Collective Phenomena*, Oxford Univ. Press 1986

B. Simon, *The Statistical Mechanics of Lattice Gases*, Vol. I, Princeton Univ. Press 1993

N.N. Bogoliubov, *Lectures on Quantum Statistics*, Vol. 2, Gordon and Breach, 1970

S.G. Brush, Rev. Mod. Phys. **39**, 883 (1967)

K. Huang, *Statistical Mechanics*, Wiley 1987

G. Gallavotti, *Statistical Mechanics: A Short Treatise*, Springer 1999

B.M. McCoy and T.T. Wu, *The Two Dimensional Ising Model*, Harvard Univ. Press 1973

T.D. Schulz, D.C. Mattis and E.H. Lieb, Rev. Mod. Phys. **36**, 856 (1964)

E. Lieb, Two-dimensional Ice and Ferroelectric Models, in *Boulder Lectures in Theoretical Physics*, Vol. XI D, K.T. Mahantappa and W.E. Brittin eds., Gordon and Breach 1969

J.B. Kogut, Rev. Mod. Phys. **51**, 659 (1979)

H.E. Stanley, *Introduction to Phase Transitions and Critical Phenomena*, Oxford Univ. Press 1974

C.J. Thompson, *Mathematical Statistical Mechanics*, Princeton Univ. Press 1972

H.M. Hugenholtz, States and Representations in Statistical Mechanics, in *Mathematics of Contemporary Physics*, R.F. Streater ed., Academic Press 1972

D. Ruelle, Helv. Phys. Acta **36**, 789 (1963)

J. Lebowitz and E. Lieb, Adv. Math. **9**, 316 (1972), Appendix by B. Simon

212   References

R. Haag, N.M. Hugenholtz and M. Winnink, Comm. Math. Phys. **5**, 215 (1967)

R. Kubo, J. Phys. Soc. Jap. **12**, 570 (1957)

P.C. Martin and J. Schwinger, Phys. Rev. **115** 1342 (1959)

J. Dixmier, *Von Neumann algebras*, North-Holland 1981

A. Leonard, Phys. Rev. **175**, 176 (1968)

J. Cannon, Comm. Math. Phys. **29**, 89 (1973)

J. Bros and D. Buchholz, Z. Phys. C-Particles and Fields **55**, 509 (1992); Nucl. Phys. **B429**, 291 (1994)

J. Goldstone, A. Salam and S. Weinberg, Phys. Rev. **127**, 965 (1962)

D. Kastler, D.W. Robinson and J.A. Swieca, Comm. Math. Phys. **2**, 108 (1966)

D. Kastler, Broken Symmetries and the Goldstone Theorem in Axiomatic Field Theory, in *Proceedings of the 1967 International Conference on Particles and Fields*, C.R. Hagen et al. eds., Interscience 1967

J.A. Swieca, Goldstone theorem and related topics, in *Cargése Lectures in Physics*, Vol. 4, D. Kastler ed., Gordon and Breach 1970

R.F. Streater, Spontaneously broken symmetries, in *Many degrees of freedom in Field Theory*, L. Streit ed., Plenum Press 1978

R.V. Lange, Phys. Rev. Lett. **14**, 3 (1965); Phys. Rev. **146**, 301 (1966)

J.A. Swieca, Comm. Math. Phys. **4** , 1 (1967)

J.D. Bjorken and S.D. Drell, *Relativistic Quantum Fields*, McGraw-Hill Book Company 1965

G. Morchio and F. Strocchi, Ann. Phys. **170**, 310 (1986)

W. Magnus, Commun. Pure Appl. Math. **7**, 649 (1954)

R.M. Wilcox, J. Math. Phys. **8**, 962 (1967)

A. Klein and B.W Lee, Phys. Rev. Lett. **12**, 266 (1964)

T.W.B. Kibble, Broken Symmetries, in *Proceedings of the Oxford International Conference on Elementary Particles*, Oxford 1965

G.S. Guralnik, C.R. Hagen and T.W.B. Kibble, Broken Symmetries and the Goldstone Theorem, in *Advances in Particle Physics*, Vol. 2., R.L. Cool, R.E. Marshak eds., Interscience 1968

R. Haag, D. Kastler and E.B. Trych-Pohlmeyer, Comm. Math. Phys. **38**,137 (1974)

L. Landau, J. Fernando Perez and W.F. Wreszinski, J. Stat. Phys. **26**, 755 (1981)

Ph. Martin, Nuovo Cim. **68**, 302 (1982)

M. Fannes, J.V. Pule' and A. Verbeure, Lett. Math. Phys. **6**, 385 (1982)

Ch. Gruber and P.A. Martin, Goldstone theorem in Statistical mechanics, in *Mathematical Problems in Theoretical Physics*, (Berlin Conference 1981), R. Schrader et al. eds., Springer 1982

W.F. Wreszinski, Fortschr. Phys. **35**, 379 (1987)

R. Requardt, J. Phys. A: Math. Gen. **13**, 1769 (1980)

G. Morchio and F. Strocchi, Ann. Phys. **185**, 241 (1988); Errata **191**, 400 (1989)

F. Strocchi and A.S. Wightman, J. Math. Phys. **15**, 2198 (1974)

G. Morchio and F. Strocchi, Ann. Inst. H. Poincaré, **A 33**, 251 (1980)

F. Strocchi, Comm. Math. Phys. **56**, 57 (1977)

A.S. Wightman, Ann. Inst. H. Poincaré, I, 403 (1964)

K. Hepp, Helv. Phys. Acta **36**, 355 (1963)

H. Ezawa and J.A. Swieca, Comm. Math. Phys. **5**, 330 (1967)

P.W. Higgs, Phys. Lett. **12**, 133 (1964) J. Bros, H. Epstein and V. Glaser, Comm. Math. Phys. **6**, 77 (1967)

R. Jost and H. Lehmann, Nuovo Cim. **5**, 1598 (1957)

F. Dyson, Phys. Rev. **110**, 1460 (1958)

H. Araki, K. Hepp and D. Ruelle, Helv. Acta Phys. **35**, 164 (1962)

A.S. Wightman, Analytic functions of several complex variables, in *Dispersion Relations and Elementary Particles*, (Les Houches Lectures), C. de Witt and R. Omnes eds., Wiley 1961

H. Araki, *Mathematical Theory of Quantum Fields*, Oxford University Press 1999

P.W. Higgs, Phys. Rev. Lett. **13**, 508 (1964); Phys. Rev. **145**, 1156 (1966); Spontaneous Symmetry breaking, in *Phenomenology of particle physics at high energy: Proc. 14th Scottish Univ. Summer School in Physics 1973*, R.L. Crawford and R. Jennings eds., Academic Press 1974, p. 529

G.S. Guralnik, C.R. Hagen and T.W. Kibble, Phys. Rev. Lett. **13**, 585 (1964)

T.W.B. Kibble, Phys. Rev. **155**, 1554 (1966)

F. Englert and R. Brout, Phys. Rev. Lett. **13**, 321 (1964).

S. Coleman, *Aspects of symmetry. Selected Erice lectures*, Cambridge Univ. Press 1985

S. Elitzur, Phys. Rev. **D 12**, 3978 (1975); G.F. De Angelis, D. De Falco and F. Guerra, Phys. Rev. **D 17**, 1624 (1978)

S. Weinberg, *The Quantum theory of Fields*, Vol. II, Sect. 21.1

R. Ferrari, L.E. Picasso and F. Strocchi, Comm. Math. Phys. **35**, 25 (1974)

J.A. Swieca, Nuovo Cim. **52A**, 242 (1967)

K. Symanzik, *Lectures on Lagrangean Quantum Field Theory*, Desy report T-71/1

G. Morchio and F. Strocchi, Removal of the infrared cutoff, seizing of the vacuum and symmetry breaking in many body and in gauge theories, invited talk at the *IX Int. Conf. on Mathematical Physics*, Swansea 1988, B. Simon et al. eds., Adam Hilger Publ. 1989

F. Strocchi, Long range dynamics and spontaneous symmetry breaking in many body systems, lectures at the Maratea Workshop on *Fractals, Quasicrystals, Knots and Algebraic Quantum Mechanics*, A. Amann et al. eds., Kluwer Academic Publ. 1988

S.S. Schweber, *An Introduction to Relativistic Quantum Field Theory*, Harper and Row 1961

G. Morchio and F. Strocchi, J. Phys. A: Math. Theor. **40**, 3173 (2007)

P.A.M. Dirac, Canad. J. Phys. **33**, 650 (1955)

O. Steinmann, *Perturbative Quantum Electrodynamics and Axiomatic Field Theory*, Springer 2000

D. Buchholz, S. Doplicher, G. Morchio. J.E. Roberts and F. Strocchi, Ann. Phys. **290**, 53 (2001).

G. Morchio and F. Strocchi, J. Math,. Phys. **44**, 5569 (2003)

W.A. Bardeen, Nucl. Phys. **B 75**, 246 (1974)

G. Morchio, D. Pierotti and F. Strocchi, Ann. Phys. **188**, 217 (1988)

R.J. Crewther, in *Field Theoretical methods in Particle Physics*, W. Rühl ed., Reidel (1980), p. 529.

Index